Practical Requirements and Exercises

Electrical Installation Series – Foundation Course

Ted Stocks

Edited by Chris Cox

MACMILLAN

First published 1999 by
MACMILLAN PRESS LTD
Houndmills, Basingstoke, Hampshire RG21 6XS
and London
Companies and representatives throughout the world

ISBN 0–333–71988–3

A catalogue record for this book is available from the
British Library.

This book is printed on paper suitable for recycling and
made from fully managed and sustained forest sources.

10   9   8   7   6   5   4   3   2   1
08   07   06   05   04   03   02   01   00   99

Printed in Great Britain by L&S Printing Co. Ltd

About this book

"Practical Requirements and Exercises" is one of a series of books published by Macmillan Press Ltd related to Electrical Installation Work. The series may be used to form part of a recognised course, for example City and Guilds Course 2360, or individual books can be used to update knowledge within particular subject areas. A complete list of titles in the series is given below.

Practical Requirements and Exercises summarises the underpinning background knowledge required for the Part I Practice.

Level 2 NVQ

Candidates who successfully complete assignments towards the City and Guilds 2360 Theory and/or Practice Part I can apply this success towards Level 2 NVQ through a process of Accreditation of Prior Learning.

Electrical Installation Series

Foundation Course
Starting Work
Procedures
Basic Science and Electronics

Supplementary title:
Practical Requirements and Exercises

Intermediate Course
The Importance of Quality
Stage 1 Design
Intermediate Science and Theory

Supplementary title:
Practical Tasks

Advanced Course
Advanced Science
Stage 2 Design
Electrical Machines
Lighting Systems
Supplying Installations

Acknowledgements

The author and publishers would like to thank the following illustration sources:

Accident Form F2508 on pp. 5 and 6 (Figures 1.5 and 1.6): "Crown copyright is reproduced with the permission of the Controller of Her Majesty's Stationery Office"; RS Components for Table 1.1 and Figures 1.39, 1.40, 1.57, 1.58, 1.59, 1.60, 1.61, 1.62, 1.64, 2.2, 2.3 and 9.11; Steve Redwood for Figure 1.42; David Wilson Homes for Figure 1.43; Farnell for Figure 9.13; Maplin Electronics for Figures 9.1, 9.12 and 9.16; F8 Imaging for Figure 1.41.; The Institution of Electrical Engineers for permission to reproduce Table 4.1 (Table 4A from the *IEE On-Site Guide*, p. 67).

Every effort has been made to trace all copyright holders, but if any have been overlooked the publishers will be pleased to make the necessary arrangements at the first opportunity.

Study guide

This studybook has been written to enable you to study the underpinning background knowledge that will be required to complete the City and Guilds 2360 Electrical Installation Practice Part I. It is likely that you will be asked to study the relevant material in advance of completing suitable practical exercises at a college, training centre or workplace under tutor supervision. The following points may help you.

☞ The practical exercises suggested in this studybook are for guidance only and your training centre/college may devise different ones according to local conditions, practices or preferences. Whether the exercises are the same or not they should all cover the same objectives and tutors will be looking for the same competences in the student. The practical work should be carried out under the supervision of a tutor in a suitable workshop. Guidance is given in this book to help you to carry out the practical work with greater understanding.

☞ Each exercise is preceded with some theory which should be studied before doing the practical work. The exercises have a check list entitled "Points to consider" which should assist you in being aware of what is expected of you. In addition you should be aware that your tutor will also be noting the following points:
  – Your ability to work with others (colleagues, supervisors and customers)
  – Your ability to treat visitors correctly
  – Your ability to contribute to your organisation's services to its customers
  – Your ability to adhere at all times to current regulations, recommendations and guidelines for health and safety

☞ A record of your achievements will be kept for you by the training centre/college and you will be able to have access to it in order that you are aware of your progress.

☞ Your safety is of paramount importance. You are expected to adhere at all times to current regulations, recommendations and guidelines for health and safety.

Your tutor may give you a programme of work. The boxes below may be used to assist you to complete your work on time.

Study times	a.m. from	to	p.m. from	to	Total
Monday					
Tuesday					
Wednesday					
Thursday					
Friday					
Saturday					
Sunday					

Programme	Date to be achieved by
Chapter 1	
Chapter 2	
Chapter 3	
Chapter 4	
Chapter 5	
Chapter 6	
Chapter 7	
Chapter 8	
Chapter 9	

Practical exercises

Below is a complete list of all the practical exercises contained in this book. The exercises have similar objectives to those in the City and Guilds Electrical Installation Practice Part 1 (Exercises 01–05). Where suggested times are given these are for guidance only.

Exercise No.	Page No.	Title
1	4	Electric shock
2	8	Fire extinguishing equipment
3	13	Safe movement of loads
4	17	Access equipment
5	18	Barriers and warning notices
6	19	Site visitors
7	20	Input services
8	21	Safety in the workplace
9	23	Tool and equipment safety
10	25	Circuit isolation
11	27	Protective clothing and equipment
12	28	Accident and emergency procedures
13	44	Connecting a 13 A plug
14	44	Connection of a junction box
15	45	Connection of a 13 A spur outlet unit and flexible cable
16	51	Wiring a 3 plate lighting circuit
17	52	Wiring a two-way and intermediate lighting circuit
18	53	Wiring a ring final circuit
19	54	Earthing and bonding
20	55	Visual inspection
21	56	Continuity of protective conductors test
22	57	Ring final circuit continuity test
23	58	Insulation resistance test
24	64	MIMS cable terminations
25	64	Terminating and fitting MIMS cable
26	65	Inspection and testing, the polarity test
27	66	Alarm circuits in MIMS cable
28	70	PVC/SWA cable termination
29	71	Methods of conductor termination
30	72	Aluminium foil sheathed cable
31	78	Connecting a 3-phase supply socket outlet
32	78	Threading and bending steel conduit
33	79	Steel conduit installation
34	79	Wiring a steel conduit installation
35	80	Complete steel conduit installation
36	85	PVC conduit bend and double set
37	85	Rigid PVC conduit for radial circuit
38	86	Radial circuit in rigid PVC conduit
39	87	Rigid PVC conduit socket installation
40	88	Wiring a rigid PVC conduit ring circuit installation
41	92	Fabricate a 90° bend in trunking
42	92	Double set in steel trunking
43	93	"T" junction in steel trunking

Exercise No.	Page No.	Title
44	94	Outside bend and conduit couplings
45	99	Fabricate a 90° bend in cable tray
46	99	"T" junction in cable tray
47	100	A reduction section in cable tray
48	101	A double set in cable tray
49	102	Cable tray combined system
50	112	Testing components
51	113	Basic soldering
52	114	Soldering components to a circuit
53	115	Replacing a component in a circuit
54	116	Terminating coaxial cable to a coaxial plug and socket
55	117	Terminating ribbon cable in an insulation displacement connection

Contents

Electrical Installation Series titles   *iii*
Acknowledgements   *iii*
Study guide   *iv*
Practical exercises   *v*
Table of contents   *vii*

1 Preparation for Work   1

Introduction   1
Attitudinal competences   1
Health and safety regulations   **1**
Electrical safety   2
Employer's responsibility   2
Employees' responsibilities   2
Electric shock   3
Electrical burns   3
Accident records   4
Fire hazards   **7**
Preventing fires   7
If a fire occurs...   7
Fire safety equipment   8
Fire extinguishers   8
Fire blankets   8
Burn injuries   8
Moving loads   **9**
Manual lifting   9
Manually handling loads   9
Guidelines   9
Manual lifting   10
Carrying long loads   10
Lifting platforms   11
Pushing and sliding   11
Assisted moving   11
Levers   11
Rollers   11
Wheelbarrows and trolleys   12
Fork-lift truck   12
Wheelstand   12
Slings and pulleys   12
The pulley block   13
Winches   13
Access equipment   **14**
Simple access equipment   14
Tower scaffold   16
Scaffolding   16
Barriers and warning notices   18
Site visitors   19
Input services   20
Safety in the workplace   21
Workplace hazards   21
Equipment hazards   22
Isolating the supply   24
Protective equipment   26

Protective clothing   26
Personal protective equipment   26
Accidents and emergencies   27

2 Basic Skills   29

The tool box   29
A basic tool box required by an electrician   29
Measuring and marking out   **30**
Transferring information to site   31
Chalked lines   31
Levels   31
Marking out   32
Drawings and diagrams   33
Block diagrams   33
Circuit diagrams   33
Wiring diagrams   33
Methods of holding work   33
Tools   **34**
Hacksaws   34
Cold chisels and bolsters   34
Files   34
Drills   34
Taps and dies   34
Hole saws   34
Reamers   35
Hole punch   35
Power tools   35
Fixings   **36**
Fixing into wood   36
Fixing into masonry   36
Fixings with other materials   36
Nailing into masonry   37
Clips and clamps   37
Pop rivet fixings   37
Cartridge fixing   37
Chasing walls and making good   37

3 PVC/PVC Cables   39

PVC/PVC sheathed wiring cable   **40**
Cable construction   40
Stripping the cable   40
Terminating the cable   41
Entry into boxes and pattresses   41
PVC insulated and sheathed flexible cable   42
Cable construction   42
Stripping the cable   42
Shaping and bending   42
Bends   42
Clipping and fixing   42
Position of cable clips   42
Circuits   **43**
Two-way lighting circuit   46

Intermediate switching circuit	46
The ring final circuit	46
Inspecting and testing	**47**
Inspection	47
Testing	47
Continuity of protective conductor test	47
Test method 1	48
Test method 2	48
Continuity of ring final circuit conductors	48
Insulation resistance	49
Polarity	50
Earth fault loop impedance	50
The test results	50

4 Metal Sheathed Cable Systems   59

Part 1
Mineral insulated cable   59

Stripping the cable	59
Terminating the cable	60
Testing	61
Polarity test	61
Connection to accessories	61
Bends	62
Cable supports	62
Alarm circuits	62
Fire alarm detectors	62
Open circuit	63
Closed circuit	63

Part 2
Steel wire armoured cable (SWA)   67

Cable construction	67
Stripping and terminating the cable	67
Bends	68
Cable supports	68
Aluminium foil sheathed cables	69

5 Steel Conduit   73

Stocks and dies	74
Conduit supports	75
Spacer bar saddle	75
The conduit system	75
Conduit bends	75
Bending conduit	75
The wooden bending block	76
Circuit protective conductor	76
Flexible steel conduit	77
Wiring lighting circuits in conduit	77
Testing steel conduit as a circuit protective conductor	77

6 Rigid PVC Conduit   81

Similarities	81
Differences	81
Bending PVC conduit	82

Terminating PVC conduit	82
Fixing PVC conduit	83
Wiring ring final circuits in conduit	84

7 Metal Trunking   89

Cutting trunking square	90
Fabricating a 90° angle	90
Fabricating a double set	90
Fabricate a "T" junction	91
Terminations into trunking	91
Cone cutter	91

8 Cable Tray   95

Fabricating cable tray   96

Fabricating a flat 90° bend producing a sharp angle	96
Fabricating a flat 90° bend producing a wide angle	96
Forming a "T" junction	96
Forming a reduction	97
Inside and outside bends	97
Fabricating an outside bend	97
Fabricating an inside bend	98
Making a cable tray bracket	98

9 Associated Electronics   103

Cables	**104**
Connections	**104**
Connectors	104
Plugs and sockets	104
Connecting to printed circuit boards	104
Mechanically fixed components	105
Connector blocks	105
Insulation displacement connection (IDC)	105
Crimp connections	106
Solder connections	**106**
Soldering irons	107
Soldering techniques	107
Preparation	107
Soldering to a pin	107
Soldering to tag strip	108
Soldering to printed circuit board	108
For good soldering joints	108
Precautions when soldering	109
Common soldering faults	109
Desoldering techniques	109
Common desoldering problems	109
Electrical measurements	**110**
Taking measurements	110
Measuring resistance	110
Analogue readings	110
Testing diodes	110
Testing transistors	111
Measuring voltage	111
Measuring current	111

1
Preparation for Work

Before undertaking any of the tasks in this studybook ensure that all safety precautions are taken, all Health and Safety guidelines are followed and all tasks carried out under supervision.

Introduction

Attitudinal competences

As part of the practical competences required on this course you will have to demonstrate your ability to work with others. You can be required to do this during any of the practical exercises that are detailed in this studybook. Whilst suggestions are given in appropriate places, your tutor may decide to put these exercises in at any time.

Basically your tutor will be asking whether you can establish and maintain effective working relationships with colleagues, customers and visitors. This will take into account such factors as how courteous you are, whether you show leadership qualities, whether you co-operate with others and can accept and respond appropriately to instructions.

Checks will be made to see how you approach health and safety matters. As part of this assessment you will be required to identify hazards, treat simulated electric shock using a dummy, organise evacuation in the event of a fire or similar emergency and so on. Most of this is likely to occur as you complete the exercises in this chapter, but to provide further evidence some may be assessed when you are undertaking any of the other practical exercises.

You may also be asked to participate in role play in order to demonstrate your ability to contribute to the improvement of an organisation's service to its customers. You could be asked to provide information and advice to customers and users of an installation. In doing so you should take into account such factors as to whether the customer is aware of the technical details of an electrical installation and selecting the best approach to give customers the information they require. Your manner and approach to the customer is vitally important. You may also be asked to identify and recommend opportunities for improving services to customers. This may involve designing a circular for a mailshot for example.

It is important to remember some general health and safety requirements before beginning the first practical exercise.

Health and safety regulations

You will find it useful to look in a library for copies of the regulations mentioned in this chapter. Read the appropriate parts and be on the look out for any amendments or updates to them.

Keeping everyone safe at work is the responsibility of the employer **AND** the employee and both are required, by law, to observe safe working practices. Various Acts of Parliament govern what employers provide in a workplace and how the employees use this provision. Some of these are briefly detailed below.

Health and Safety at Work Etc. Act 1974

This Act applies to everyone who is at work and it sets out what is required of both employers and employees. The aim of this Act is to improve or maintain the standards of Health, Safety and Welfare of all those at work.

Figure 1.1   *The employer will provide, and the employee should wear, eye protection when grinding materials.*

Under the Health & Safety at Work etc. Act 1974 are other Regulations and Codes of Practice including:

Management of Health & Safety at Work Regulations 1992
The Electricity at Work Regulations 1989
Manual Handling Operations Regulations 1992

There are other laws and regulations which deal with Health, Safety and Welfare at Work. Some of those with reference to the electrical industry could include:

The Factories Act 1961
Safety Representatives and Safety Committees 1977
Notification of Accidents and General Occurrences Regulations 1980
Control of Substances Hazardous to Health Regulations 1988 (COSHH)

We shall briefly consider the requirements related to electrical safety and the responsibility of employers and employees.

Electrical safety

Accidents involving the use of electricity may result in electric shock, fire, burns or injury from mechanical movement.

Accidents can occur as a result of:
- faulty materials or equipment
- poorly maintained or misused equipment
- misbehaviour or carelessness
- protective devices and equipment which are not used, or are used incorrectly

All these factors apply to electrical accidents.

Strict observance of the applicable laws, standards and codes of practice is imperative to reduce the possibility of accidents.

Laws which must be observed with particular regard to electricity include:
- The Health and Safety at Work etc. Act 1974
- The Electrical Supply Regulations 1988 and
- The Electricity at Work Regulations 1989

- BS 7671 Requirements for Electrical Installations, the Regulations published by the Institution of Electrical Engineers, are not law but are accepted as standard practice for electrical installation work.

The British Standards Institution (BSI) also produce Standards and Codes of Practice. Any which are applicable to the work activity should be complied with.

These Laws, Standards and Codes of Practice relate to the manufacture, installation and use of electrical equipment. Before working on electrical equipment all staff should be properly qualified, trained and competent in their work.

Employer's responsibility

Employers are required to provide and maintain a working environment for their employees which is, as far as practicable, safe and without risk to health.

Factors to consider regarding the safe working environment include:
- access in areas such as corridors, staircases and fire exits (Figure 1.2)
- maintenance of a reasonable working temperature
- fume and dust control
- maintenance of adequate ventilation
- all areas must also be suitably and adequately illuminated
- facilities for washing, sanitation must be supplied
- supply of first-aid equipment.
- provision and maintenance of suitable safe tools and equipment for use by employees
- any training in the use of such equipment that is necessary
- any information or supervision as may be required
- a safe working method
- supply of protective equipment and protective clothing where required
- storage, handling and transporting of goods

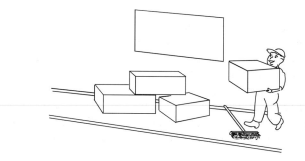

Figure 1.2   Access areas should be kept clear

Employees' responsibilities

An **EMPLOYEE** can be prosecuted for breaking safety laws.

Employees are required by law to:
- take reasonable care for their own health and safety and not to endanger others
- co-operate with their employer on health and safety procedures
- not interfere with tools, equipment etc. provided for their health, safety and welfare
- correctly use all work items provided in accordance with instructions and training given to them

The first practical exercise to consider is regarding the basic procedures to be taken in the event of electric shock.

Electric shock

If a person comes into contact with a phase (live) conductor while they are also in contact with earth, current will pass through their body. It can also happen if a person touches two live conductors (Figure 1.3) as their body will complete the circuit between them. In either case the person concerned will have received an electric shock by "direct contact".

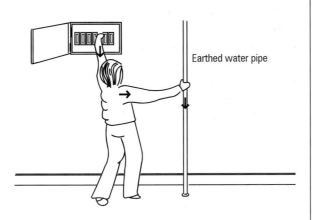

Earthed water pipe

Figure 1.3   Electric shock from direct contact

Electric shock may be caused by the metal case of an electrical appliance becoming "live". This may occur during a fault to earth on any part of the installation. At this time all the "exposed metal parts of the electrical installation" (exposed conductive parts) become live. Then anyone touching the casing and other metalwork could become "a conductor" and suffer an electric shock by "indirect contact". This situation lasts until the protective device disconnects the circuit.

If you find someone who has received an electric shock your first priority must be to take care not to become a casualty yourself. For this reason there is a correct procedure that should be followed when dealing with a person who has received an electric shock.

FIRST the connection of the person to the electrical supply must be broken without causing any further injury:

IF possible cut off the electricity supply

IF this is not possible the casualty will have to be pulled clear **BUT** only with a dry insulator ("dry" because water conducts electricity). A dry insulator could be rubber gloves, a newspaper, a rope or a wooden pole (Figure 1.4). Alternatively, standing on a rubber or plastic mat or dry wood could prevent you receiving an electric shock. Do **NOT** touch the casualty's bare skin without taking precautions to prevent electric shock.

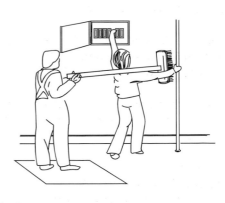

Figure 1.4   Pull casualty away with non-conductive material.

Once disconnected from the supply:

IF the casualty is not breathing start resuscitation immediately and call for expert help.

IF the casualty is unconscious, but breathing, place them in the recovery position (Figure 1.5) and get help. They may be burned or bleeding as a result of the shock or fall and require further medical aid.

Figure 1.5   Recovery position

Electrical burns

High currents can create arcs that may cause serious burns. With high current the voltage may be very low so the electric shock potential is slight.

A good example of how a high current, low voltage electrical burn can be caused is the misuse of a lead acid battery as used in a car. The voltage is only 12 V d.c. but if the battery is shorted out the current can be over 100 A. A spanner carelessly laid down or dropped across the terminals of such a battery will cause an arc. This produces considerable heat and may cause serious burns.

If you or your colleagues receive an electrical burn get expert medical help immediately.

Accident records

Following **ANY** accident employees need to note the following details in order to fill in the accident record which **must** be kept in any workplace:

- the date and time of the accident or dangerous occurrence
- the place where the accident or dangerous occurrence took place
- a brief description of the circumstances
- the name of any injured persons
- the sex of the injured persons
- the age of the injured persons
- their occupations
- the nature of the injuries

There may be other details required by particular organisations. These may range from witnesses to suggestions for the prevention of further accidents.

Major injuries will have to be reported and a sample of the accident report form required by the Health and Safety Executive for major injuries is shown on the next two pages (Figures 1.6 and 1.7). You will notice that on the sample accident report form there is a space for accidents where one of the factors involved is the electricity supply cable, wiring, apparatus or equipment.

You may find it useful at some stage to attend a recognised first aid course.

1
Electric shock

Objective: This task requires you to practice basic procedures in the event of electric shock.

NOTE: NO live parts or equipment are required to achieve the objective of this exercise. The artificial respiration and resuscitation techniques must only be carried out using recognised methods.

Procedures:

UNDER COMPETENT PROFESSIONAL SUPERVISION carry out the procedures for electric shock which should include:

- a demonstration of how a "casualty" should be removed from contact with "live" parts
- resuscitation
- heart massage
- laying in the recovery position
- completion of an accident report form in the approved manner

You may also be required to demonstrate your ability to work with others during this exercise.

Points to consider:

- was the electricity supply cut off?
- were the professional emergency services notified?
- was the model removed, if required, from contact in an acceptable manner?
- was the model placed in the correct position for artificial resuscitation?
- was resuscitation carried out correctly?
- was the model placed in the correct recovery position?
- was the accident form completed in the approved manner?

Health and Safety at Work etc Act 1974
The Reporting of Injuries, Diseases and Dangerous Occurrences Regulations 1995

HSE
Health & Safety
Executive

Report of an injury or dangerous occurrence

Filling in this form
This form must be filled in by an employer or other responsible person.

Part A

About you

1 What is your full name?

2 What is your job title?

3 What is your telephone number?

About your organisation

4 What is the name of your organisation?

5 What is its address and postcode?

6 What type of work does the organisation do?

Part B

About the incident

1 On what date did the incident happen?

/ /

2 At what time did the incident happen?
(Please use the 24-hour clock   )

3 Did the incident happen   ve address?

Yes ☐ Go to ques

No ☐ Where   e incident   en?

☐ else   your   anisation – give the
name,   postcode

☐ at someone   remises – give the name,
      ress and po

☐ in a public place – g   etails of where it
      ha

If you do not   postcode, what is
the name of the local authority?

4 In which department, or where on the premises,
did the incident happen?

Part C

About the injured person

If you are reporting a dangerous occurrence, go
to Part F.
If more than one person was injured
please attach the details asked for in Pa   Part D for
each injured person.

1 What is their full name?

2 What is their home ad   ode?

3 Wh   e p   numb

      old are

5 Are they

☐ male?

☐   ale?

6 Wha   eir job title?

_ was the injured person (tick only one box)

☐ one of your employees?

☐ on a training scheme? Give details:

☐ on work experience?

☐ employed by someone else? Give details of the
employer:

☐ self-employed and at work?

☐ a member of the public?

Part D

About the injury

1 What was the injury? (eg fracture, laceration)

2 What part of the body was injured?

F2508 (01/96)

Continued overleaf

Figure 1.6   Accident form F2508 (front). Crown copyright is reproduced with the permission of the Controller of Her Majesty's Stationery Office.

5

3 Was the injury (tick the one box that applies)

- [ ] a fatality?
- [ ] a major injury or condition? (see accompanying notes)
- [ ] an injury to an employee or self-employed person which prevented them doing their normal work for more than 3 days?
- [ ] an injury to a member of the public which meant they had to be taken from the scene of the accident to a hospital for treatment?

4 Did the injured person (tick all the boxes that apply)

- [ ] become unconscious?
- [ ] need resuscitation?
- [ ] remain in hospital for more than 24 hours?
- [ ] none of the above.

Part E

About the kind of accident

Please tick the one box that best describes what happened, then go to Part G.

- [ ] Contact with moving machinery or material being machined
- [ ] Hit by a moving, flying or falling object
- [ ] Hit by a moving vehicle
- [ ] Hit something fixed or stationary

- [ ] Injured while handling, lifting or carrying
- [ ] Slipped, tripped or fell on the same level
- [ ] Fell from a height

    How high was the fall?

    | | metres |

- [ ] Trapped by something collapsing

- [ ] Drowned or asphyxiated
- [ ] Exposed to, or in contact with, a harmful substance
- [ ] Exposed to fire
- [ ] Exposed to an explosion

- [ ] Contact with electricity or an electrical discharge
- [ ] Injured by an animal
- [ ] Physically assaulted by a person

- [ ] Another kind of accident (describe it in Part G)

Part F

Dangerous occurrences

Enter the number of the dangerous occurrence you are reporting. (The numbers are given in the Regulations and in the notes which accompany this form)

| |

Part G

Describing what happened

Give as much detail as you can. For instance

- the name of any substance involved
- the name and type of any machine involved
- the events that led to the incident
- the part played by any people.

If it was a personal injury, give details of what the person was doing. Describe any action that has since been taken to prevent a similar incident. Use a separate piece of paper if you need to.

Part H

Your signature

Signature

| |

Date

| / / |

Where to send the form

Please send it to the Enforcing Authority for the place where it happened. If you do not know the Enforcing Authority, send it to the nearest HSE office.

For official use		
Client number	Location number	Event number

[ ] INV REP [ ] Y [ ] N

Figure 1.7   Accident form F2508 (back). Crown copyright is reproduced with the permission of the Controller of Her Majesty's Stationery Office.

Fire hazards

Fire requires fuel, oxygen and heat. If all three are not available then a fire will not start. If a fire has started, then by removing any one of the three the fire will be extinguished. For example the use of a foam fire extinguisher cuts off the supply of oxygen. Alternatively if there is no more fuel, combustible material, then a fire will go out.

Enough heat can be produced during initial combustion to maintain or accelerate the fire, resulting in rapid spreading of the fire.

Fires can be started in numerous ways such as people smoking carelessly, friction heat, sparks, naked flames, fuel leaks and faults, failures in equipment resulting in overheating and chemical reactions.

Preventing fires

There are precautions that we can take to prevent fires from starting, many of which are common sense. For example we should take care to store combustible materials away from heat sources, and all areas should be kept free from combustible rubbish and dust.

Equipment which could cause a fire should be regularly maintained and serviced, including electrical equipment and associated cables, plugs and flexes. Fuel pipes should be checked for leaks and flammable materials should be suitably stored at controlled temperatures.

If a fire occurs...

Make sure that you understand the hazards involved and know what to do in the event of a fire.

Places of work are required to have a fire procedure and if a fire occurs these should be followed immediately.

This will generally involve

- raising the alarm
- calling the fire service
- where appropriate using the correct fire fighting equipment for the particular type of fire
- shutting down equipment if possible
- clearing personnel from the area to the fire assembly point
- shutting all doors when evacuating the premises to prevent the spread of fire and smoke and to limit the extent of damage to property
- reporting to the supervisor at the prearranged assembly point

Table 1.1

Type of fire	Water (red panel)	Foam (beige panel)	Dry powder (blue panel)		CO$_2$ gas (black)
			Dry powder to BS 5423	ABC Dry powder to BSEN 3 1996	
Class A Paper, wood, textiles	✔	✔		✔	
Class B Flammable liquids such as oil, paint, petrol, paraffin, grease		✔	✔	✔	✔
Class C Flammable gases such as LPG, butane, propane, methane			✔	✔	✔
Electrical hazards			✔	✔	✔

The fire drill procedure that is displayed in your place of work must always be followed. Make sure you are aware of the emergency procedures which cover evacuation in the event of other emergencies, for example in the case of an explosion, contaminated fumes or terrorist activity.

Regulations covering fire safety include
- Health and Safety at Work etc. Act 1974
- Fire Precautions Act 1971 and Regulations
- Building Regulations 1976
- Fire Services Act 1947

Fire safety equipment

It is important to find out the location of all available fire fighting equipment and the types of fire for which it can be used.

Fire extinguishers

All British made fire extinguishers are colour coded to indicate their particular purposes, Table 1.1 shows the different types and main uses. From the 1st January 1997 all new certified fire extinguishers, under BS EN 3, must have red bodies. In the U.K. colour identification panels are placed on or above the operating instructions. Fire extinguishers with full body colour codings may still be found and can continue to be used until they need replacing. Halon extinguishers, colour code green, are being phased out of service due to the adverse effects they have on the environment.

Fire blankets

Fire resistant blankets are used to cut off the air supply to extinguish the fire. These are generally used on fat or oil fires, for example when a deep fat fryer catches fire, and they can also be used to smother clothing fires.

Burn injuries

If you are to deal with minor burns you should first wash your hands and then flush the burn with plenty of clean, cool water. Do not try to remove clothing that may be sticking to the burn but apply a sterilised dressing. If the burn is serious, get expert medical help immediately.

2
Fire extinguishing equipment

Objective: This task requires you to select and demonstrate the correct use of fire extinguishing equipment.

Procedures:

NOTE: Actual fires are not required and fire extinguishers need not be actually set off to achieve the objectives of these assignments.

Under tutor supervision carry out any of the tasks below or those identified by your tutor.

- Select the correct fire extinguisher to deal with a fire inside a mains electrical control unit and demonstrate the procedure for dealing with such a fire. Explain the reasons for your choice.
- Select the correct fire extinguisher to deal with a fire which has broken out in a pile of waste paper and cardboard boxes and demonstrate the procedure for dealing with such a fire. Explain the reasons for your choice.
- Select the correct fire fighting equipment to deal with a fire which has broken out in a chip pan in the kitchen of a factory. Demonstrate the procedure for dealing with such a fire. Explain the reasons for your choice.
- Assume that a large fire has broken out in one room of the industrial site where you are working. You are first on the scene. Demonstrate the correct procedures to follow in these circumstances.

Points to consider:

- were the approved procedures observed promptly?
- was limitation of damage to persons observed as a priority over limitation of damage to property?
- were the professional emergency services notified in the event of an emergency or potential emergency?
- was the fire alarm system activated promptly?
- were approved procedures in the event of emergency warnings observed?
- was any damage to property following the emergency identified and was any appropriate action taken to minimise further damage?

Moving loads

Manual lifting

A load is an object which has to be moved or lifted.
It could be that 100 m reels of 1.5 mm^2 cable have to be moved from the floor on to a work-bench (Figure 1.8).

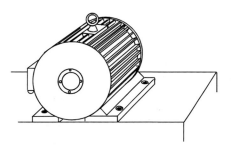

Figure 1.8

Another example is a heavy electric motor (Figure 1.9) which has to be moved from the stores into the workshop area.

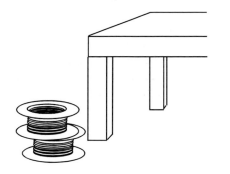

Figure 1.9

Both the above examples require a load to be moved from one place to another but the methods used to achieve this would be very different.

Manually handling loads

It is important to recognise that moving a load can involve a number of different considerations. The Manual Handling Operations Regulations 1992, which came into force on January 1st 1993, recognise the possible risks involved in moving loads. By following the guidance in the flow charts and check lists shown in the Guidance on the Regulations when carrying out manual handling operations it is possible to limit the risk of injury. Many manual handling injuries arise from repeatedly using the wrong technique or incorrect posture. This can even lead to permanent disability.

If it is necessary to move a load we should first assess the task and reduce the risk of injury to the lowest reasonably practicable level. This could require the use of mechanical assistance.

Loads over 20 kg generally need lifting gear so remember – if it is appropriate to use equipment, it has been provided and you have been trained to use it – then it is a **LEGAL** requirement for you to do so.

Guidelines

In general there are a number of factors that must be considered when a load has to be moved which include:

Moving the load may create a risk in, for example reaching up or reaching with a twisting movement. If excessive distances are involved there should be sufficient rest periods.

Figure 1.10

The weight of the load could cause an injury so it will be necessary to find out how heavy it is. There may be documentation with the load or a label on it or maybe the weight of the load can be estimated by looking at it.

A bulky or difficult shape may be difficult to grip. Some objects may be greasy or wrapped in loose packing and these will need extra care. Some objects have got suitable handling points where you can get hold of them. Care must also be taken if they have any sharp corners or are hot.

Figure 1.11

When carrying heavy loads the route and destination should be checked and cleared of all obstacles (Figure 1.12). Check there is enough headroom, or whether it will be necessary to stoop, and make sure that the floor is not slippery or uneven. Check that there is enough light to see where you are going and if you are carrying a bulky light load make sure that there will be no sudden gusts of wind.

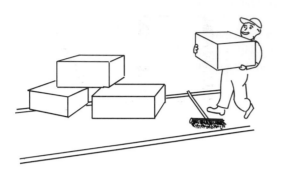

Figure 1.12

Personal protective equipment or other clothing should not be used unless absolutely necessary. Wearing gloves may impair dexterity although they may be required because of cold conditions or to avoid injury from rough or sharp edges. Other protective clothing may make free movement difficult.

After making an assessment of the task a decision must be made as to how the work is to be safely carried out. This may involve more than one person, using special equipment, preparing the working area or using safety clothing. Remember if your task requires the use of appropriate equipment, this has been provided by your employer and you have been trained to use it, then it is a legal requirement for you to do so.

Manual lifting
Whenever a load is to be lifted careful consideration needs to be given to the task and if it is to be lifted manually extra consideration needs to be given to the position of the body.

It is important to keep the spine in its naturally upright position (Figure 1.13). If the back is unnaturally arched forward there is a greater risk of injury to it. The knees should be bent so that when the legs are straightened, the load is lifted. To achieve this the load should be kept close to the body with arms as straight as possible and both hands should be used.

To allow for the centre of gravity of the load the body should lean back slightly when lifting allowing the body to act as a counter balance to the load.

When the load has been lifted and the body is straight care must be taken to avoid sudden turning or twisting as this can also cause damage to the back.

Naturally upright

Figure 1.13

Carrying long loads
Long loads being carried by a single person may cause injuries to either the person carrying the long load or somebody else in the vicinity! The person carrying the load should ensure that the centre of gravity is directly related to the carrying position.

In Figure 1.14 the centre of gravity of the ladder is approximately above the shoulder on which it is being carried. To avoid unnecessary injury to others the front end of the load is kept high.

Long ladders being carried any distance should be carried on the shoulders of two people, one at each end.

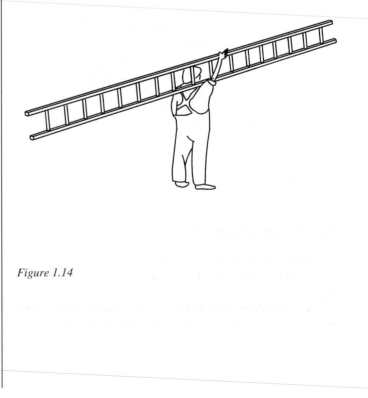

Figure 1.14

Lifting platforms

When a load has to be lifted for stacking or lifted on to the shoulders to be carried, it is useful to have a lifting platform so that the lift can be done in two stages (Figure 1.15). This will reduce the risk of strain or injury.

Figure 1.15   Lifting platform

Pushing and sliding

Lifting a load generally requires more effort than moving a load in a horizontal plane, so it is easier to push or slide a load than to lift it (Figure 1.16). If a load is to be pushed or pulled care must be taken not to damage the operative's back. To work at maximum efficiency with the least possibility of damage to the person, the back should be kept straight and the legs should do the pushing or pulling.

Figure 1.16   Pushing or sliding a load

Assisted moving

Pushing or pulling a load may be easier than lifting, but a heavy load on a flat surface can create a large friction resistance area.

This friction can be reduced if rollers are used between the load and the floor.

Levers

So that rollers can be placed under the load it has to be lifted one end at a time. The lifting can best be carried out with a lever. This is placed under one side of the load and then pushed down (Figures 1.17 and 1.18). As the part of the lever that is pushed down on is a great deal longer than the end that is lifting the load a mechanical advantage is created.

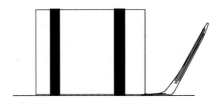

Figure 1.17

Figure 1.18

Rollers

When the load has been raised rollers can be placed underneath as shown in Figure 1.19. The load can then be pushed gently forward on two or three rollers and further rollers placed under the front end as necessary.

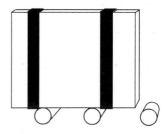

Figure 1.19

Wheelbarrows and trolleys

A **wheelbarrow** (Figure 1.20) can be used to carry heavy or bulky loads. Load the barrow so that it will not overbalance, use both hands, keep a straight back and steady the barrow before moving off.

Figure 1.20   A wheelbarrow

A **sack barrow** (Figure 1.21) has two wheels and is more stable than a wheelbarrow. It is still very necessary to load the barrow correctly to avoid overbalance.

Figure 1.21   A sack trolley or hand truck

A **flat trolley** (Figure 1.22) has four wheels and it is often used in stores where materials are constantly being moved. When on level ground a flat trolley is usually pulled whereas barrows are usually pushed. It may be necessary to prevent trolleys from moving at the wrong time in which case chocks (blocks or wedges) need to be placed to prevent the wheels turning.

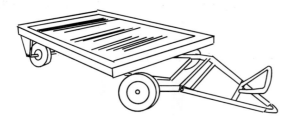

Figure 1.22   A flat trolley

Fork-lift truck

Another way of moving a load is by using a fork-lift truck. These are often used in large stores where the goods are to be stacked on pallets. Only authorised and trained personnel are allowed to use these.

Wheelstand

A **wheelstand** (Figure 1.23) is a raised platform on two wheels and two feet to avoid the need for lifting the load.

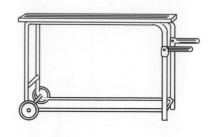

Figure 1.23   A wheelstand

Slings and pulleys

When a heavy load has to be lifted vertically slings and pulleys can be used (Figure 1.24), but care must always be taken to ensure that the supports are capable of taking the maximum load. The maximum Safe Working Load (SWL) should **NEVER** be exceeded. Hands and fingers should be kept clear of the slings as they could become trapped if the slings should move when the load is raised.

After the load has been lifted care must be taken to make sure that nobody can get trapped or crushed by the load.

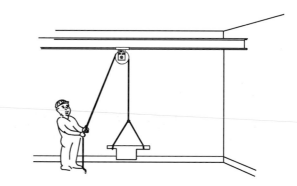

Figure 1.24   Pulleys can help to lift larger loads

NEVER leave a suspended load unsupervised. When you lower a load, do it gently into position and, before you remove the lifting equipment, make sure that the load is stable and will not topple over. Every load has a centre of gravity which will not always be the centre of the object.

Many objects that have to be lifted have sharp edges and corners and rope slings should not be used unless these are protected as they may become damaged.

The pulley block

The pulley block consists of a continuous chain or rope passing over a number of pulley wheels as shown in Figure 1.25. Pulley blocks should be regularly tested and their safe working load displayed on them which should **NEVER** be exceeded.

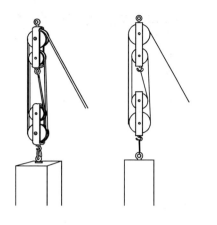

Figure 1.25   Pulley block

When loads are suspended on pulley systems they have a tendency to swing and twist. This problem is often overcome by having a stabilising or control rope tied to the load and manned by a person given the sole responsibility of keeping the load straight.

Loads should never be left suspended in mid-air without someone to watch them. The area under the load should be kept clear at all times in case the load should fall. When the load has been lowered gently into position it should be checked for movement before the sling is taken away. Where a pulley system is used the amount of effort required is only half that of a one pulley system. A four pulley system only requires a quarter of the effort of a one pulley system.

Winches

A simple winch, as in Figure 1.26, consisting of a drum around which a rope is wound, can also be used to raise loads. A crank handle rotates the drum and takes up or lets out the rope thus raising or lowering the load.

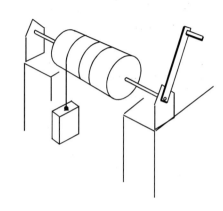

Figure 1.26   Winch

3
Safe movement of loads

Objective: Demonstrate safe manual and mechanical movement of loads.

Under tutor supervision, and with help if required, carry out either the assignments below or those set you by your tutor.

- Using the correct methods lift a load of about 15 kg from the floor and place it on a bench.
- Carry a load of about 15 kg from the bench top to a new position about 5 m away and place it on the floor.

Points to consider:

- was any attempt made to find out how heavy the load was?
- were any possible sources of hazard identified?
- were possible sources of hazard dealt with in the appropriate manner before the assignments were attempted?
- was the correct stance used to lift the object?
- was the object held in a safe way?
- when the object was carried was the back kept straight?
- when the object was placed on the ground was the balance maintained?
- were current regulations, recommendations and guidelines for Health and Safety observed at all times?

Access equipment

Simple access equipment

Access equipment provides a means of reaching the area where work has to be carried out. Different access equipment will be required depending on the height at which work is to be carried out. Reaching just above shoulder height may only require a step-up whereas scaffolding may be required for access to work on the roof of a building.

There are some basic rules which apply whichever access equipment is required.

- All access equipment should be set up on a firm level base.
- The equipment chosen must be suitable for the task so that the user does not have to overreach.
- All access equipment should be inspected regularly to ensure that it is in good condition. This does not just mean whether or not it is broken but also looking to see if the surface is slippery because of mud or ice, or other similar hazard.

To reach up a short distance in comfort use a very simple piece of equipment called a **step-up**, often referred to as a hop-up (Figure 1.27).

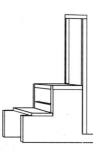

Figure 1.27   A step-up or hop-up

For work up to the height of a ceiling a **pair of steps** (Figure 1.28) could be used. These should be high enough for the user to stand with his knees below the top of the steps, which should be open to their fullest extent and should be set on level and firm flooring.

Figure 1.28   A pair of steps

Some work at ceiling height involves working over some distance and in such cases **trestles and a platform** may be used (Figure 1.29). When using trestles open the "A" frames to their full extent and if stay bars are fitted lock them into place.

Figure 1.29

The platform can be either two scaffold planks wide (at least 450 mm) or a lightweight staging and it should be placed no higher than two thirds of the way up the trestles. The platform must not overhang the trestles by more than four times the thickness of the boards (Figure 1.30) and the minimum overhang allowed is 51 mm. So if two scaffold planks are used, each 50 mm thick, then they must not overhang by more than

$$4 \times 50 = 200 \text{ mm}$$

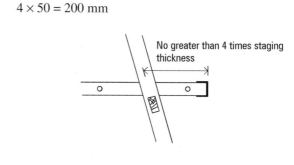

No greater than 4 times staging thickness

Figure 1.30

Due to the extent of injuries that could result from falling the maximum height for the platform must not exceed **4.57 m**. If the platform is over 2 m high a pair of steps should be used to gain access. The platform and trestles should always be dismantled before moving them to a new position.

If access to somewhere higher is required then a **ladder** should be used. It is very important when working on a ladder to remember the rule about standing it on firm level ground (Figure 1.31).

Figure 1.31   Ensure the ladder stands on firm level ground

There are also a number of other points to remember:

- Ladders must not be too short as this could cause the user to over-reach and fall off.
- Ladders must extend no less than 5 rungs or 1.05 m above the working platform unless there is an adequate handhold to reduce the risk of overbalancing (Figure 1.32).

Ladders must extend no less than 5 rungs or 1.05 m above the working platform.

Figure 1.32

- Ladders must be inspected frequently to ensure they are in good working order and that no rungs are damaged, missing or slippery. Damaged ladders should be withdrawn from use, and clearly labelled that they are damaged and not to be used.

When using **extension ladders** there must be an overlap of rungs (Figure 1.33):

2 rungs for ladders with a closed length of up to 5 m.
3 rungs for ladders with a closed length of up to 6 m.
4 rungs for ladders with a closed length of over 6 m.

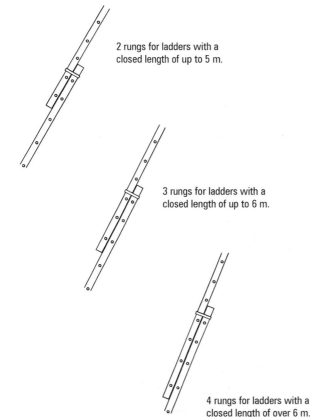

2 rungs for ladders with a closed length of up to 5 m.

3 rungs for ladders with a closed length of up to 6 m.

4 rungs for ladders with a closed length of over 6 m.

Figure 1.33

When moving ladders any distance they should be carried on the shoulders of two people, one at each end of the ladder (Figure 1.34).

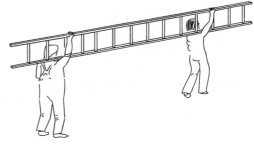

Figure 1.34

Ladders should be raised with the sections closed. Two people may be required to raise the heavier ladders. One stands with one foot on the bottom rung and holds the stiles to steady the ladder while the second person stands at the top and raises the ladder above their head and walks towards the bottom moving their hands down the ladder as they go.

Figure 1.35

When the ladder is in position it must follow the 4:1 rule. This means that the height of the ladder above the ground must be 4 times the distance the ladder is out from the foot of the wall. The ladder will then be at an angle of 75° with the ground.

If the ladder is over 3 m long then it must be secured or another person must "foot" the ladder. To secure the ladder it must be lashed to a secure position like a scaffold pole. A drainpipe or gutter is not a secure position and should not be used. In some cases it is necessary to secure ladders of less than 3 m as even short falls can cause injuries.

Figure 1.36   *A ladder lashed to a scaffold pole.*

To foot the ladder another person must stand with one foot on the bottom rung, one foot on the ground and both hands holding the stiles (Figure 1.37). The purpose of footing the ladder is to prevent it moving and the person carrying out this task must remain alert and observant at all times. Footing a ladder is therefore only recommended when the activity is likely to be of short duration and lashing is impractical.

Figure 1.37

Tower scaffold

For working at, and not just access to, these higher levels a tower scaffold (Figure 1.38) can be used. Training in the erection, use and inspection of the tower scaffold should be given. Operatives who have not been given this training should **NOT** erect or use the scaffold without suitable supervision.

The minimum base measurement for any tower is 1.21 m and for outside use the height must **NOT** exceed 3 times the smallest base measurement. If it is higher than 4 m outriggers must be fitted and if it exceeds 9 m it must be anchored to the ground or tied to the building. The overall height of the tower must not exceed 12 m unless it has been specially designed to do so.

The castors at the foot of the tower must be locked before anyone climbs the tower, and whilst anyone is on the tower it must not be moved. When the tower has to be moved it should only be moved by pushing at the base.

Any platforms used for working from must have toe-boards and guardrails for the safety of those working on and those below. Precautions must also be taken to ensure that access cannot be gained to incomplete scaffold towers and that only authorised persons can gain access at any time.

Where safety equipment, for example a harness, is used on tower scaffold it must be designed and constructed to prevent serious injury in the event of a fall.

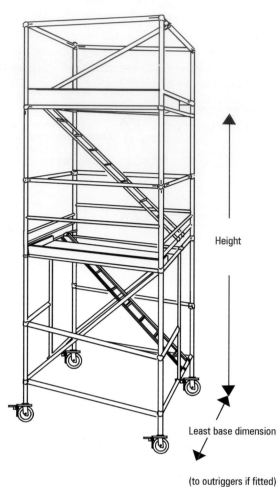

Height

Least base dimension

(to outriggers if fitted)

Figure 1.38

Scaffolding

Full scaffolding is another alternative but this should only be erected by a competent scaffolder.

4
Access equipment

Objectives: Examine, carry, erect and secure a 3 m extension ladder. Select and use appropriate access equipment.

Under tutor supervision, and with help if required, carry out either the assignments below or those set you by your tutor.

- Carry out a visual examination of a ladder and report on its condition.
- With help, if required, carry a 3 m ladder a distance of 10 m using a safe method
- With help, if necessary, erect and secure a 3 m ladder to an acceptable standard.
- Select, erect and dismantle the most appropriate access equipment to carry out a simple installation task given to you by your tutor.

Points to consider:

- were instructions accepted from the supervisor and responded to appropriately?
- were requests from colleagues accepted and responded to appropriately?
- if new colleagues were present was appropriate support and information provided?
- was the task carried out in a courteous manner throughout?
- were any defects in the ladder reported?
- was the route/erection area checked for any possible hazards?
- were any hazards reported and recorded?
- was the ladder lifted and carried safely?
- was the ladder lifted into position correctly?
- was the slope of the ladder to an acceptable gradient?
- did the top of the ladder extend the required amount?
- was the ladder ascended in a safe manner?
- was the ladder secured in a safe acceptable manner?
- was the most suitable access equipment selected for the tutor led assignment?
- was the equipment erected and dismantled in the correct manner?
- were current regulations, recommendations and guidelines for Health and Safety observed at all times?

Barriers and warning notices

Where it is possible for a person to fall more than 2 m into an excavation or hole then barriers **MUST** be erected (Figure 1.39), but it is safe practice to erect them around **ANY** excavation. This will not only prevent anyone falling into the excavation or hole but also prevent vehicles or materials getting too close. Warning notices (Figure 1.40) to control access to the site may also be required.

Figure 1.40   Warning notices

Figure 1.39   Barriers

5
Barriers and warning notices

Objective: Use barriers and warning notices to control access.

Under tutor supervision carry out either the assignment below or one set you by your tutor.

* Ensure that access to the work site is controlled by using barriers and warning notices.

You may also be required to deal with customers and visitors to the site and your attitude to this will be monitored.

Points to consider:

* were the correct barriers chosen to control access to the work site?
* were warning notices placed prominently?
* were current regulations, recommendations and guidelines for Health and Safety observed at all times?

Site visitors

Employers are responsible for ensuring that visitors to the site (Figure 1.41) are included when drawing up their safety policy.

You will remember that authorised site visitors should be asked to identify themselves and identify by name the person they wish to see. Then they should sign in, as a record has to be kept of all personnel on site and when they arrived and left. They must also be made aware of the safety procedures they need to follow while they are on site. These instructions must be given clearly and it must be ensured that they have been understood. Visitors should have safety equipment appropriate to the site conditions. If these are not available then no access to the site should be permitted.

Figure 1.41   Site visitors

6
Site visitors

Objective: To ensure visitors are dealt with according to agreed procedures.

Under tutor supervision carry out either the assignment below or one set you by your tutor.

*   Assume that you are responsible for a visitor to the site on which you are working. What is the procedure that you should follow?

Points to consider:

*   was the visitor asked for some form of identity?
*   was the visitor asked to identify, by name, who they were visiting?
*   was the visitor signed in?
*   was the visitor given any safety equipment?
*   was the visitor escorted appropriately?
*   was the visitor escorted off the site at the end of the visit?
*   was the visitor treated with courtesy and in the correct manner to promote goodwill and good practice?
*   were the requirements of the Health and Safety at Work Act 1974 observed?

Input services

The input services to any site, whether it be home, office, factory or building site, need to be located and identified. Examples of electricity and gas input services are shown in Figures 1.42 and 1.43.

Figure 1.42   Commercial and industrial gas input services

Figure 1.43   Domestic gas and electrical input services

7
Input services

Objective: To locate and determine the suitability of the input services.

Under tutor supervision carry out either the assignment below or one set you by your tutor.

- Locate the input services for the area indicated by your tutor.
- Identify the electricity, gas and water input services.
- Verify the suitability of the input services for the intended purpose.

You may be observed as to your attitude in accepting instructions and as to how you respond. Also how you deal with the situation should you not get clear instructions on how you should proceed.

Points to consider:

- were the input services located correctly?
- was the electricity input service identified?
- was the gas input service identified?
- was the water input service identified?
- were the gas and water suitable for the intended purpose?
- were the requirements of the Health and Safety at Work Act 1974 observed?

Safety in the workplace

Workplace hazards

Employees should look out for hazards in the workplace and report them so that they do not cause an accident. The workplace itself should be maintained in good order and in good repair. If a potentially dangerous situation is noticed, such as an insecure structure, it should be reported so that it can either be remedied immediately or fenced off so that access is prevented.

In any workplace floors and work-benches should be kept clear of rubbish and any tools and equipment should be put away when they are not needed. Caps should be replaced on bottles (Figure 1.44), and if any liquids are spilled they should be cleaned up immediately. Waste materials should be removed or stored in suitable containers. Access routes must be kept clear at all times (Figure 1.45) and temporary hazards, such as where a floor board has been removed, should be adequately guarded.

Figure 1.44   *Replace caps on bottles*

Figure 1.45   *Keep access routes clear*

Machinery should not be interfered with; in particular, it is extremely dangerous to remove guards. A machine must not be used by someone who has not been suitably trained and given authority to operate the equipment. Long hair should be tied back and loose clothing should not be worn as it may get caught in machinery.

Your place of work could be an environment which has sources of hazard which need to be considered before starting work. For example you may be asked to work in an area where others are drilling walls or using toxic substances, using lifting equipment, digging trenches and so on. Your work may involve wiring under floors or in roofs where access may be limited.

Other possible sources of hazard to be considered include the risk of drowning, being hit by falling objects or whether there is a lack of suitable lighting in the working area.

When work has to be done in an area that may be hazardous consideration must be given as to whether the workplace can be made safer or whether there are some hazards that can be removed or reduced. Barriers may need to be erected, protective equipment worn or training undertaken.

8
Safety in the workplace

Objective: To ensure safety in the workplace before work commences.

Under tutor supervision carry out either the assignment below or one set you by your tutor.

* Your working environment is to be in a chill store and an adjoining room. In the adjoining room other trades have been stripping the paint off prior to redecorating. Assess possible sources of hazard and prepare the worksite to ensure that it is suitable for work to commence.

Points to consider:

* were any possible sources of hazard noted?
* was the worksite prepared to acceptable levels of risk?
* were the requirements of the current regulations, recommendations and guidelines for health and safety observed?

Equipment hazards

Deficiencies in equipment should be reported as soon as they are noticed. For example a broken ladder rung (Figure 1.46), failed lighting or a frayed sling on a pulley and sling system are hazards that should be reported before they cause injury to a user.

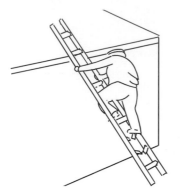

Figure 1.46

For any job the correct tool should be used and it should be in a safe condition. Tools which show signs of damage or wear can be potentially dangerous and should not be used until repaired or replaced (Figure 1.47).

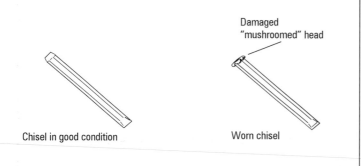

Chisel in good condition

Damaged "mushroomed" head

Worn chisel

Figure 1.47

Protective clothing and other equipment should be worn where it is necessary. Eye protection should be worn when grinding metals, cutting or chasing and hard hats should be worn on all building sites. Protective clothing and equipment should conform, where necessary, to the mandatory standards.

Printed rules, warning notices and instructional posters should be read. These should be prominently displayed and other information, instruction and training may be given, which is equally important and must be followed. It is irresponsible being given this training to not practice it constantly.

Safety signs give warning of particular dangers and will show the particular type of protection required for the conditions concerned.

When portable electrical equipment is used in the workplace or on construction sites, accidents can often be prevented by following a few simple rules.

A visual check on cables and plugs, which are particularly liable to damage, can prevent a serious accident (Figure 1.48).

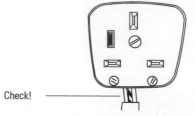

Check!

Figure 1.48   Visually check cable and plugs

Electrical equipment may develop faults which do not affect its operation, yet equipment may present a potential hazard. Regularly testing the equipment (Figure 1.49) and implementing repairs help to ensure accidents are prevented.

Figure 1.49   Testing equipment before use

In general all electrical equipment should be connected to earth through a circuit protective conductor. However, if the equipment conforms to the appropriate standards it may be classified as Class II equipment, often called double insulated. The "classes" of equipment are described in BS 2754:1976 "Construction of electrical equipment for protection against electric shock" which is is based on IEC Report 536. The symbol in Figure 1.50 would be displayed on the case of equipment which is Class II and in these situations no circuit protective conductor should be used.

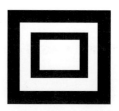

Figure 1.50   British Standard symbol for Class II (double insulated) equipment

Where equipment is used out of doors or in damp environments it should have a residual current device (RCD; Figure 1.51) protecting the circuit with an operating current of 30 mA. This means that should a fault develop between phase and earth or neutral and earth the circuit would automatically switch off before the fault current could reach 30 mA.

On construction sites or in factories it is advisable to limit the voltage used on handheld portable equipment to 110 V a.c. This is supplied through a transformer which limits the line voltage to 55 V above earth potential. This is generally regarded as the maximum safe working voltage for a.c. On d.c. the maximum safe working voltage is 120 V.

Figure 1.51   RCD

9
Tool and equipment safety

Objective: To ensure the safety of tools and equipment.

Under tutor supervision carry out either the assignments below or ones set you by your tutor.

- A chisel shows signs of wear as illustrated in the diagram. What should be carried out to make this safe?

- A report on an electric drill states the following:
  The drill was found to have a two core flexible cable. The plug on the end was a 13 A type with a 10 A fuse.
  Drill label:

Voltage 110 V	
Current 2.0 A	
Power 200 W	

State what action you would take given this information.

Points to consider:

- were any possible sources of hazard noted?
- was the worksite prepared to acceptable levels of risk?
- were the requirements of the Health and Safety at Work Act 1974 observed?

Isolating the supply

The Electricity at Work Regulations must be followed when isolating the supply.

Before any circuit is to be worked on it should, whenever possible, be isolated. Many accidents have occurred over the years where the wrong circuit has been cut off. It is therefore very important that all circuits and isolators are clearly labelled as to what they control and circuits are tested to confirm that they are isolated before they are worked on. It is important that an isolator cuts off an electrical installation, or any part of it, from every source of electrical energy.

The device for cutting off the supply must not only be suitable for operation in abnormal or fault conditions but also be placed in an accessible location. Where switches are used as isolators a clear air gap must exist between the contacts so that they cannot accidentally reconnect. They must also be clearly marked so that there is no doubt whether the switch is ON or OFF.

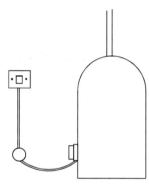

Figure 1.52   Immersion heater (double pole switch)

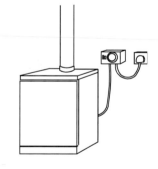

Figure 1.53   Domestic boiler (plug and socket)

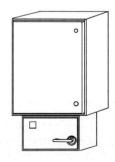

Figure 1.54   Distribution board (isolator)

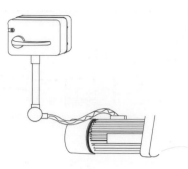

Figure 1.55   Motor (triple pole isolator)

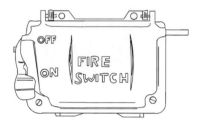

Figure 1.56   Fireman's switch (isolator)

When equipment has been isolated for working on there are a number of points that may need consideration:
- all exposed electrical connections should be tested to see if they are dead
- the test equipment used must be proved before and after use

It is necessary to post notices at the isolator warning that the equipment is being worked on and the switch or isolator should be **LOCKED** in the **OFF** position.

If it becomes essential to work on or near live equipment a permit to work may be necessary. This ensures that both the person working on the live equipment and a person in authority know of the risks that are being taken. The important factor is **DO NOT WORK ON LIVE EQUIPMENT** without first examining every other possibility. If it is then found necessary to work live, every safety precaution must be considered and taken to prevent danger. The correct test equipment must be used. Test probes which conform to the Health and Safety Executive Guidance Note GS 38 are recommended.

Where the electrical equipment has its own source of supply, such as batteries, capacitors or generators, then it is important that, as these cannot be isolated within themselves, all possible precautions are taken to prevent a dangerous situation occurring.

10
Circuit isolation

Objective: To ensure that circuits to be worked on are isolated.

WARNING!
This test must be carried out under competent supervision

UNDER TUTOR SUPERVISION carry out either the assignment below or one set you by your tutor.

- Inspect and isolate a suitable source of supply.
- Confirm isolation using proprietary voltage test equipment which has been tested for correct working both before and after the test.

Equipment:

- A suitable source of supply
- Proprietary voltage test equipment

Points to consider:

- how was the source of supply isolated?
- were any warning notices required?
- was it necessary to LOCK OFF the circuit?
- have any visual defects with the test equipment been identified?
- has the correct working of the test equipment been verified (i) before the test and (ii) after the test?
- has the circuit been identified as "live" or "dead"?
- were the requirements of the Health and Safety at Work Act 1974 observed?

WARNING!
This exercise should be carried out in accordance with the Electricity at Work Regulations 1989 and all live conductors should be shielded and insulated from touch. All test probes should meet the requirements of the Health and Safety Executive Guidance Note GS38.

Protective equipment

We have already looked at some of the hazards that should be considered on the worksite. Now we will look more closely at the types of protective clothing and equipment available that are worn in the interests of health and safety.

Protective clothing and equipment needs to be kept in good repair, stored appropriately and replaced when no longer in good condition. Such clothing and equipment should fit the wearer properly and training should be provided to ensure that the wearer understands why the equipment needs to be used and any risks involved.

Protective clothing

This includes items such as gloves (Figure 1.57), safety helmets (Figure 1.58), protective footwear (Figure 1.59), protective clothing for adverse weather (Figure 1.60), coveralls and aprons.

Leather gloves can safeguard against cuts from manual handling of heavy and sharp objects and gloves may be required when working outside in very cold conditions. If there is a risk of an electric shock, electric arc or burns wearing suitable gloves can provide the protection required.

Figure 1.57   Electrician's gloves

Head protection may be in the form of a safety helmet where there is a risk of injury from falling or flying objects.

Figure 1.58   Head protection

Protective footwear should be worn where there is a risk from manual or mechanical handling, electrical work or any work done in hot or cold weather conditions. When choosing footwear consideration should be given to grip, resistance to water or hazardous substances, flexibility and comfort for the wearer.

Figure 1.59   Footwear protection

Protective clothing should provide protection for the body against such things as extremes of temperature, chemical splashing or harsh weather conditions. If the work environment is in, say, a chemical works or a chill store, appropriate protective clothing should be selected including such items as thermal socks for warmth. If work is to be undertaken that requires high visibility in hazardous conditions then reflective workwear should be worn.

Figure 1.60   Thermal socks

Personal protective equipment

This includes such items as safety goggles, dust masks and ear muffs.

Eye protection (Figure 1.61) may be required in dusty environments or where chemicals could be splashed. In such conditions dust masks may also be required.

Figure 1.61   Eye protection

Ear protection (Figure 1.62) should be used in very noisy environments such as on a site where there are pneumatic drills or power presses or when using cartridge-operated tools. As a general rule if you have to shout to be heard by someone who is only 2 metres away then ear protection should be used.

Figure 1.62   Ear protection

> ### Remember
> **Eye protection** may be required in dusty environments and in such conditions dust masks may also be required.
>
> **Ear protection** is required where there is a high level of background noise, which may not be due to the activities you are involved in.

Safety signs (Figure 1.63) will usually be found on construction sites where it is necessary to wear protective clothing. These are mandatory signs circular in shape, white symbol on a blue background and they show what **MUST** be done. Other areas may not have signs and possible dangers must be assessed by the worker and appropriate protective gear worn.

Figure 1.63   Safety signs

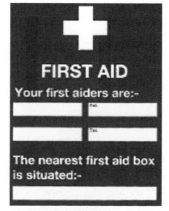

11
Protective clothing and equipment

Objective: To use appropriate protective equipment.

Under tutor supervision carry out either one of the assignments below or others set you by your tutor.

Assess any potential risks at the site and select and use appropriate protective clothing and equipment where your working environment is:

- in a foundry
- in a boiler room
- in a sawmill
- outside during the cold and wet winter months

Points to consider:

- were appropriate possible risks noted?
- was appropriate protective clothing and equipment selected?
- was appropriate protective clothing and equipment used correctly?
- were the requirements of the Health and Safety at Work Act 1974 observed?

Accidents and emergencies

Remember:

> Accidents are **unplanned**.
> Accidents do not **"just happen"**.
> Accidents are **caused by people**.

An **"ACCIDENT"** is an unexpected or unintentional event which is often harmful.

There are three main causes:
- people who become unsafe because of such factors as boredom, horseplay, carelessness or lack of knowledge
- people who have provided or maintained an unsafe environment
- people who have misused a safe environment

Accidents can be prevented by observing the following:

Employees should take reasonable care of their own health and safety and that of others. This means that they will be expected to do their job in a sensible way and they should not run around or fool about. They should also make sure that they are fit for

work and not overtired, ill or suffering from the effects of drink or drugs.

Always report an accident, whether it results in injury or not, as it may prevent another more serious accident from the same cause.

It is important to note where the first-aid box is located (Figure 1.64) (this is identified by a white cross on a green background) and to find out who are the appointed first-aiders.

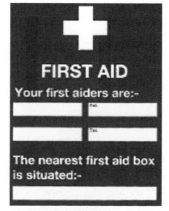

Figure 1.64   First aid location sign

There are other guides available from the HSE, industry and workplaces that can help in understanding the employees' responsibilities.

Just as employers are responsible for providing training, information and supervision, so the employees are responsible for carrying out the work in the manner in which they have been trained. These points are important: learn them off by heart.

To prevent an accident:
- keep the work area clean and tidy (clear up the rubbish!)
- take care near machines (don't interfere with safety guards and be careful not to have long loose hair or clothing)
- take care of your tools (keep them in good condition)
- wear protective clothing or equipment when it is necessary.
- take note of any rules, regulations or safety signs and obey them
- take reasonable care for your own health and safety and do not endanger others
- look out for hazards and report them (do not assume that this has already been done)
- always report an accident and note down the details required to fill in the accident form
- know where the first-aid box is located and who is the appointed first-aider

Employers must ensure that, in the event of an accident, however slight, the details are recorded in a register kept at the workplace. The employer may be required to report these accidents to the Health and Safety Executive, who may wish to investigate further. They may also inspect the accident register at the same time.

Major injuries that result in fractured bones, loss of sight or hospitalisation for more than 24 hours must be recorded in the accident register. In addition every accident involving an employee must be notified to the Health and Safety Executive if the employee is unable to work for three or more consecutive days. This is recorded on the form on pages 5 and 6 .

An emergency is any event which requires immediate action and in the event of an emergency your priority is to limit injury to persons before limiting damage to property.

You should:
- raise the alarm
- notify the professional emergency services
- suspend work immediately
- isolate equipment from its power source if it is safe to do so
- then proceed in accordance with safety procedures to a recognised assembly point.

Remember

An accident involves personal injury whilst an emergency may involve risk to health, life or property.

12
Accident and emergency procedures

Objective: To demonstrate prompt and effective application of accident and emergency procedures.

Under tutor supervision carry out either the assignments below or others set you by your tutor.

What would you do when on being the first to enter the workshop one morning you notice:
- a smell of burning and thick smoke or
- a recorded message on the answerphone telling you that a bomb has been planted nearby that will explode in 5 minutes or
- toxic fumes coming in the window from a neighbouring chemical factory

Points to consider:

- were approved procedures observed promptly?
- was damage limited to persons before damage limited to equipment or property?
- were professional emergency services notified?
- were alarm/alert/evacuation/disaster safety systems activated?
- were approved procedures in the event of an accident/emergency warning observed?
- was damage to property identified following the accident/emergency and was appropriate action taken to minimise further damage?
- were agreed procedures observed in the event of an emergency, if appropriate?
- were current regulations, recommendations and guidelines for Health and Safety observed at all times?
- was the accident reported promptly and recorded in the approved format?

2

Basic Skills

There are no practical exercises contained within this chapter but it should provide you with some background information on the basic skills you will require before completing the exercises in the remainder of this book. You will, of course, be advised of the correct procedures by your tutor and should regard this chapter as providing you with a reminder of some of the relevant points. It could be helpful for you to add other comments in your own words to act as an aide memoire in the future.

On completion of this chapter you should be able to:

◆ select appropriate tools and equipment for work in different situations
◆ state the procedure required for fixing into vertical surfaces constructed of cement rendered brick/blockwork
◆ take measurements using scale rules
◆ recognise the instruments needed to take measurements on round objects
◆ identify marking out equipment for defined situations
◆ transfer information from drawings
◆ identify block, circuit and wiring diagrams

The tool box

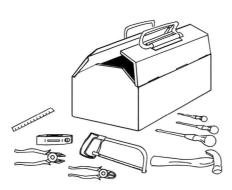

Figure 2.1

A basic tool kit required by an electrician

A basic tool kit, which should built up by all trainee electricians, would include such items as

- pin hammer
- claw hammer
- screwdrivers for slotted head and cross head screws
- junior hacksaw
- spanners
- spirit level
- plumb-bob and line
- metric rule
- bolster cold chisel
- club hammer
- knife (not Stanley type)
- side cutters
- strippers
- electrician's terminal screwdrivers for lighting and power connections
- pliers
- small file

As your career progresses you will obviously acquire more tools to meet your needs. You should purchase a secure toolbox to put your own hand tools in. Remember that damaged tools must be repaired or replaced for your own and others' safety.

Tools for testing the cable and specialist items, such as conduit benders, may be provided by your employer.

Tools are expensive to replace. Take care of your own and your employer's tools, keep them clean and in good condition and it is also be a good idea to consider insuring them.

Measuring and marking out

Taking measurements can be carried out using many different devices (Figures 2.2 and 2.3). The selection of the correct one may depend on factors such as; the accuracy required, the shape of the object and the positioning among many others.

Figure 2.2   *Rules and measuring tapes*

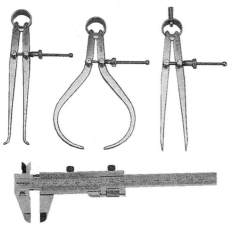

Figure 2.3   *Callipers and Vernier scale*

Rules are by far the most commonly used devices for measuring. However there are probably more different types of rule in use than any other measuring instrument. The length of a rule can vary from a few centimetres to several metres long. The graduations can also vary depending on what the rule was designed for. Figure 2.4 shows some of the more common scale rules in use.

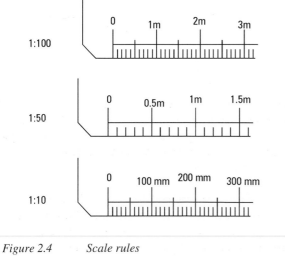

Figure 2.4   *Scale rules*

Tapes are used for measuring longer distances. A tape consists of a flexible measuring strip which is kept rolled up in a case (Figure 2.5).

Figure 2.5   *Pocket and long tapes*

Calipers (Figures 2.6 and 2.7) can be used to measure inside and outside pipes, conductors or other difficult to measure shapes. Measurements are taken by adjusting the leg tips of the calipers until they just touch the inside or outside faces of the pipe and then withdrawing the calipers taking great care not to change the position of the legs.

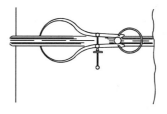

Figure 2.6   *Outside callipers*

The caliper legs are then placed on a rule and the dimension determined.

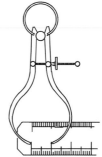

Figure 2.7   *Using a rule to take a measurement*

Some calipers now have a dial from which the measurement can be read directly.

Very accurate readings, to hundredths of a millimetre, can be taken using a micrometer.

It is often necessary to check if work is "square". To do this requires the use of a try square (Figure 2.8). The stock of an engineers try square is placed against the true edge of the work to confirm whether the work is square.

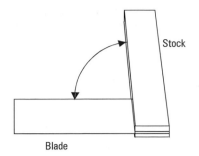

Stock

Blade

Figure 2.8   A square

Transferring information to site

Electrical designers usually detail the position of electrical accessories and equipment on scaled drawings. This means that a comparatively small sheet of paper can represent a large area with a lot of electrical equipment in it. To transfer the information from the drawing to the actual area the scale of the drawing must be known. Then, using a scaled rule on the drawing, as in Figure 2.9, the actual measurement can be determined. A tape can then be used on site to measure the exact position as shown in Figure 2.10.

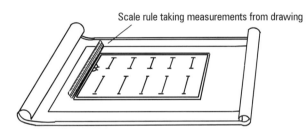

Scale rule taking measurements from drawing

Figure 2.9   A scaled drawing

Figure 2.10 shows how measurements can be transferred to wall positions but it is not always so easy when lights have to be marked out, particularly as a one-man operation.

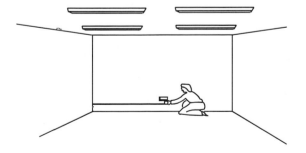

Figure 2.10   Measurements being transferred on site

One of the best ways to overcome the problems of holding tapes at ceiling level is to mark out the lighting points on the floor. Using a plumb line the point on the floor can be transferred to the ceiling (Figure 2.11).

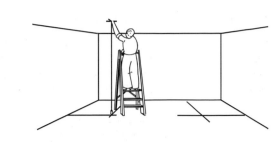

Figure 2.11

Chalked lines

Where long straight lines are required the "chalked" line (Figure 2.12) can be used. The line should be covered in chalk and with at least two pencil or chalk marks on the wall and with each end of the line anchored it can be "pinged" so that the chalk is transferred to the surface. This method can be used for vertical or long horizontal runs. The plumb line may be used in this manner providing the line is covered in chalk.

Chalked line

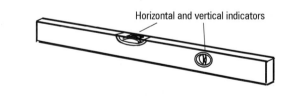

Figure 2.12

Levels

In some circumstances it is not possible to use this "chalk line" method of marking out. In such cases a spirit level (Figure 2.13) and straight edge have to be used. The spirit level is a very helpful aid when installing any vertical or horizontal equipment, but it must be looked after. If the edges of it become damaged or the spirit tube becomes loose, the accuracy of the level will be affected.

Horizontal and vertical indicators

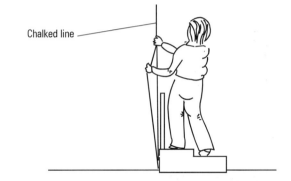

Figure 2.13   Spirit level

Water levels (Figure 2.14) are used to transfer a horizontal level from one location to another. This device is based on the principle of water finding its own level. Using this technique it is possible to transfer a level from one room to another or around corners.

Decide where the first horizontal line is to be drawn using data from drawings and using a rule from either the floor or ceiling. Make a single mark on the wall at the correct height. Check that there are no kinks in the apparatus and get someone to hold one of the calibrated sight tubes until the middle mark is on a level with the mark on the wall. Open the screw top of the sight tube. Take the second tube to where it is required and raise, or lower, the tube carefully until the level of the water in the first sight tube is exactly in line with the mark on the wall again. Hold both of the tubes steady, check the levels again, and when they are both the same make a second mark in line with the water level in the second tube. Do not make any unnecessary marks on the wall, and if the room is decorated use chalk which can be erased when the work is completed.

Figure 2.14   Water level

Marking out

When marking out on different surfaces the correct tools should be used. If a cross has to be marked on a brick wall then chalk or crayon can be ideal. However if an accurate mark has to be made on a metal surface then a scriber (Figure 2.15) and centre punch (Figure 2.17) may be required.

A square, as in Figure 2.16, is used for measuring and checking angles of 90°, it comprises two parts, the stock and the blade and it is very accurate.

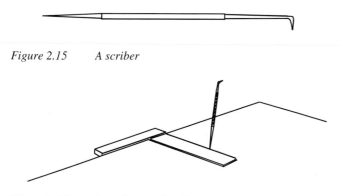

Figure 2.15   A scriber

Figure 2.16   A scriber used with a square on a metal surface

A scriber can also be used with a straight edge, a heavy duty steel or alloy strip along which the scriber can be drawn to mark metal or other surfaces. Incline the scriber away from the straight edge and in the direction of movement to give a clean line.

Metal surfaces that are to be drilled should first have the position marked out with a scriber. Then a centre punch should be used to create an indentation that the drill will locate in. This will reduce the possibility of the drill sliding across the smooth metal surface and ensure that the hole is drilled in the correct position.

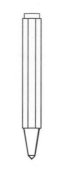

Figure 2.17   A centre punch

The electrician is often going to be involved with working on metal surfaces. These surfaces may be parts of distribution boards and bus-bar chambers, or on installation equipment such as trunking and cable tray.

Taking a section of trunking as an example, changes in direction will often have to be fabricated on site. This means that in addition to the normal site tools, equipment such as squares, scribers and centre punches are also required. This fabrication to fit a site requirement can be quite complex and if care is not taken costly material can be wasted. It is sometimes a good idea to chalk out the shape of the finished bend on the floor or some other surface which will not be damaged (Figure 2.18). If this is carried out carefully, and to scale, the measurements can be transferred directly to the trunking by laying the trunking onto the drawing.

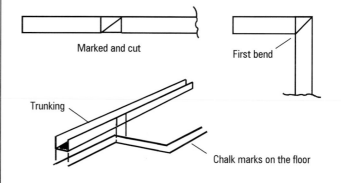

Figure 2.18

A similar process may be used when conduit shapes are to be made but compasses may be necessary to get the exact shape required (Figure 2.19).

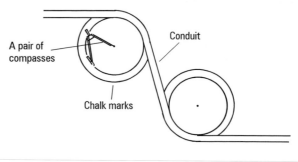

Figure 2.19

Drawings and diagrams

The most frequently used diagrams that you should be able to recognise are:

- block diagrams
- circuit diagrams
- wiring diagrams

Block diagrams

These indicate the sequence of components or equipment. Each item is represented by a labelled block, as in Figure 2.20.

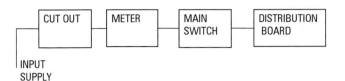

Figure 2.20   Block diagram

Circuit diagrams

Circuit diagrams (Figure 2.21) are used to show the operation of a circuit and the circuit components. The symbols represent pieces of equipment or apparatus and the diagram will show how the circuit works.

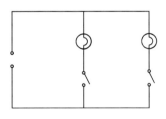

Figure 2.21   Circuit diagram

Wiring diagrams

Wiring diagrams (Figure 2.22) indicate the locations of the components in relation to one another, and actual cable connections, and are more detailed than circuit diagrams.

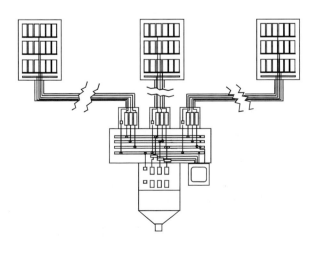

Figure 2.22   Wiring diagram

Methods of holding work

There are many different types of vice (Figure 2.23) for holding work. The vice is usually attached securely to a work bench or support stand (for example a conduit vice).

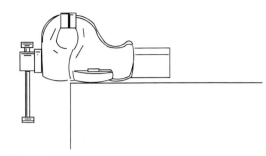

Figure 2.23   A vice

Clamps (Figure 2.24) can be used when third hands are required. "One hand" grip clamps are now more common.

Figure 2.24   A "G" clamp

When round objects such as conduit needs to be held securely vee blocks (Figure 2.25) or a purpose-made pipe vice can be used.

Figure 2.25   Vee block

There are many different types of workbenches (Figure 2.26), some of which are portable and provide a strong bench surface. These either incorporate a vice or can generally be fitted with one.

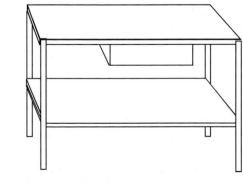

Figure 2.26   Workbench

Tools

Hacksaws

Work to be cut with a hacksaw (Figure 2.27) needs to be held firmly in a vice and the line of cut should be close to, but clear of, the jaws. A hacksaw can be used with either a vertical or horizontal blade. The blade can be turned by release of the tension adjusters and rotating the blade support blocks.

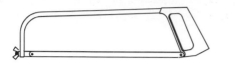

Figure 2.27   Hacksaw
Remember – the blade teeth should always face forward.

Cold chisels and bolsters

When using a cold chisel (Figure 2.28) always ensure that the chisel head is safe to use. Grind away any mushrooming metal remembering to use eye protectors during this process.

Bolster

Cold chisel

Figure 2.28   Cold chisel and bolster

Files

There are many variations of file face for filing away rough edges on such items as trunking or cable tray (Figure 2.29). A handle should always be fitted to the file before use and the work piece should be held securely in a vice.

Figure 2.29   File (with handle)

Drills

Drills (Figure 2.30) need to be chosen for their intended use. There are special drills for cutting masonry or hardened steel and a wide variety for general purpose drilling. They should all be stored carefully and kept sharp and ready for use.

Figure 2.30   Drill

Taps and dies

For cutting internal threads a tap is used with a tap wrench to turn the tap to cut the thread (Figure 2.31). When cutting the thread rotate the tap one full turn clockwise then a half turn anticlockwise. this will break off the swarf that has been produced. The use of cutting compound will extend the life of the tools and make cutting threads and holes easier.

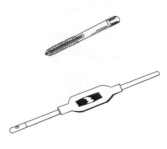

Figure 2.31   Tap and wrench

For cutting external threads, such as on the outside of conduit, a stock and die is used (Figure 2.32). Threads are cut using the dies which are held in the stocks (more details can be found in Chapter 5).

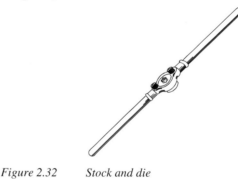

Figure 2.32   Stock and die

Hole saws

Hole saws (Figure 2.33) are often used with electric drills but the speed must be kept low to prevent the blade from burning up and cutting compound should be used.

Figure 2.33   Hole saw

Reamers

A conduit reamer (Figure 2.34) is a hardened steel cone shaped swarf cutter and the size of the cut hole regulates the depth the reamer is allowed to go.

Figure 2.34   Conduit reamer

Hole punch

A hole punch (Figure 2.35) is only suitable for use on sheet metal. A pilot hole is drilled first and the punch and die assembled on each side of the material to be pierced. The draw bolt is passed through the die and screwed into the punch. The draw bolt is then tightened up with the keywrench exerting a shearing force on the cutter which pierces the hole.

Figure 2.35   Hole punch

Power tools

So far we have considered hand tools but powered machines are often used such as an electric drill (Figure 2.36), hacksaw or grinder.

Figure 2.36   Cordless electric drill

Types of building materials

Table 2.1 shows some of the more common materials and typical fixings suitable for the type of material used.

Table 2.1

Material Type	Fixings	Suitable Tools
Brickwork Common or facing bricks are made from clay. They are fairly easy to drill and have good fixing properties. Engineering bricks are harder and fairly difficult to drill.	Screw fixings Fibre or plastic plugs Masonry nail	Hand drill Electric drill Rotary hammer drill Hammer
Concrete Made from cement and aggregate in dense and lightweight forms as for concrete blocks **Concrete Blocks** Lightweight concrete blocks are easy to drill and have fair fixing properties. Dense concrete blocks are fairly easy to drill and have good fixing properties.	Screw fixings Fibre or plastic plugs Bolts – several different methods Resin anchors Stud anchors	Hand Drill Electric drill Rotary hammer drill
Steel	Snap on fixings, girder clips Shot fired cartridges	Hammer, screwdriver and spanner Cartridge powered fixing tool
Plasterboard Lath and plaster Hardboard, plywood, fibre board and chipboard Difficult to get good fixings	Spring toggle fasteners Gravity toggle Cavity sleeve Cavity wall anchor Dry lining boxes	Drill Screwdriver
Wood Easy to drill Good fixing properties	Wood screws Nails	Drill Screwdriver Hammer

Fixings

Fixing into wood

Nails (Figure 2.37) are often used for wood fixings. Generally oval wire nails are used for replacing skirting boards or architraves and round wire nails for fixing straps in down drops. Galvanised clout nails should be used where moisture can affect the nails. Many of the fixings supplied now come with the correct type and size of nail but generally where fixings have been provided pre-drilled use the same diameter nail as the hole. Nails should not be held in the mouth but in a suitable container, such as a nail pouch.

Figure 2.37   Wire nail

Holes drilled into wood can be made using either a hand drill or a power drill.

There are many types of wood screw available (Figure 2.38) including
- countersunk – for general woodwork
- cross head screws – must be used with a special screwdriver
- coach screw – this has a square head so that it can be tightened with a spanner
- security screw – this type of screw once screwed in cannot be unscrewed

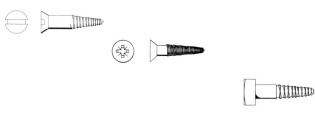

Figure 2.38   Countersunk, cross head and coach screws

Before screwing a fixture onto wood mark out the position of the holes by centering a pencil or bradawl through the fixture to leave a mark on the wood. A bradawl can then be used to make a fixing hole for small screws or a drill for the larger sizes. The screw should then be placed in the hole and partially screwed in. Remove the screw and fully locate the fixture in the correct position. Insert the screw through the fixture fixing hole and tighten the screw. If the fixture has more than one hole fix the uppermost hole first and allow the fixture to hang from this fixing screw whilst the other fixing screws are located and tightened. Always ensure that fixtures are secure and correctly positioned.

Fixing into masonry

Drills used for masonry have cutting faces formed in hard carbide. It is better to use a low speed when drilling masonry to avoid overheating the drill as this can cause wear to the cutting face of the drill.

When fixing screws into masonry a plastic plug (Figure 2.39) is generally used.

Figure 2.39   Plastic plug

A development of the plastic plug is the hammer-in screw (Figure 2.40), which is suitable for securing conduit, trunking, battens and small enclosures into masonry.

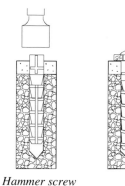

Figure 2.40   Hammer screw

Fixings with other materials

Fixings for concrete, soft brick and other similar materials also include anchor bolts (Figure 2.41), wedge anchors and chemical anchors. These latter anchors can be used particularly into difficult materials, where fixings have to be made near outside edges or in groups with close spacings. Full instructions as to their usage is usually given on the packet or box.

Figure 2.41   Expanding anchor

Fixings for plasterboard, hardboard and other sheet materials the fixings include screw plugs, anchors, toggle and cavity fixings (Figure 2.42).

Figure 2.42   Cavity wall anchor

Nailing into masonry

Special hardened masonry nails can be used for fixing small items such as surface cable clips and wooden battens. Goggles or appropriate eye protection must be used when driving in masonry nails.

Clips and clamps

Clips and clamps (Figure 2.43) are used on building structures where drilling is out of the question. They are very easy to fix and can save both time and labour.

Figure 2.43   Clamp fittings

Pop rivet fixings

Pop rivets (Figure 2.44) are normally used for fixing two pieces of metal sheet together. Fixing is obtained by placing the rivet in a pre-drilled hole and squeezing together using a rivet gun.

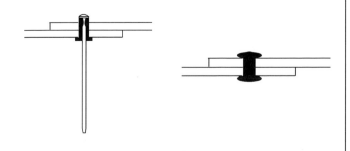

Figure 2.44   Pop rivet

Cartridge fixing

Making fixings into steel, masonry, concrete and brickwork can be achieved much quicker using a cartridge tool, however they should never be used by untrained operators.

Chasing walls and making good

Wiring systems may need to be concealed in a chase cut in brickwork or in plaster. Parallel lines should be drawn on the wall the width of the chase to be cut and the size of any terminal boxes that are to be fitted (Figure 2.45). As a certain amount of debris will be made the floor should be covered with a protective sheet. Suitable personal protective equipment should be worn to prevent injury to eyes and/or hands.

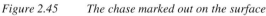

Figure 2.45   The chase marked out on the surface

A wide flat chisel should be tapped with a hammer to outline the chase and the plaster should be cut away to expose the brickwork underneath. Check at this stage whether or not the cables and mechanical protection can be positioned in the chase with a covering of at least 5 mm depth of covering plaster. If it is necessary to chase into the brickwork to achieve the required depth then use a masonry drill to drill a honeycomb of holes over the whole area (Figure 2.46). Then use a cold chisel and hammer with short sharp taps to remove the necessary brickwork. Check that the chase is of the required depth.

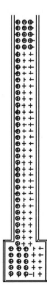

Figure 2.46

The chase will have to be made good once the wiring is completed by filling with plaster.

Figure 2.47   Ensure terminal boxes are level and plumb

Add your own notes here.

3

PVC/PVC Cables

Your tutor will suggest tasks to cover the practical competences as detailed in the practical exercises shown in the following chapters. The basic principles will be the same in that you will be required to prove your practical ability in installing electrical systems and equipment and to supplement this with relevant underpinning knowledge.

You will also be required to practically demonstrate your ability to work with others and the suggested scenarios in the Introduction to this book, or others suggested by your tutor, will be used in conjunction with the practical exercises in order to make this assessment.

On completion of this chapter you should be able to:

◆ strip a PVC insulated and sheathed cable without damage to the insulation or conductors
◆ terminate cables into electrical accessories to an acceptable standard
◆ wire a 3 plate lighting circuit
◆ wire a two-way and intermediate switched lighting circuit
◆ wire a ring final circuit
◆ inspect the above circuits against the listed criteria
◆ test the above circuits and tabulate the results

The tool kit required

For stripping the cable
• knife (not Stanley type)
• side cutters
• strippers
in addition, for terminating the cable
• terminal screwdriver for lighting connections
• terminal screwdriver for power connections
• pliers
• small file
in addition, for fixing the cable
• pin hammer
• screwdriver for No. 8 and No. 10 screws
• junior hacksaw
in addition, for testing the cable
• low reading ohmmeter
• 500 V Megohmmeter
• earth fault loop impedance tester

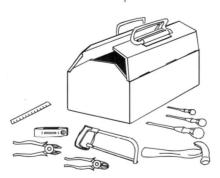

Figure 3.1

PVC/PVC sheathed wiring cable

Cable construction

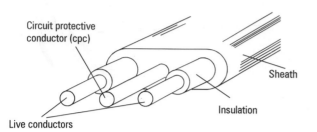

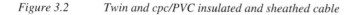

Figure 3.2   *Twin and cpc/PVC insulated and sheathed cable*

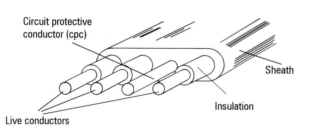

Figure 3.3   *Three-core and cpc/PVC insulated and sheathed cable*

Stripping the cable

As you can see (Figures 3.2 and 3.3), the circuit protective conductor in this type of cable is uninsulated, so this can be used as a guide for the knife when removing the cable sheath.

Allow the knife to lay against the cpc (Figure 3.4) and push slowly with a slight sawing movement. When you have reached the required point peel the sheath back and cut it off with side cutters.

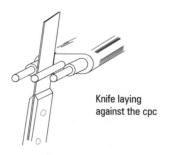

Knife laying against the cpc

Figure 3.4   *Care must be taken NOT to cut into your hand*

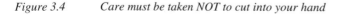

Many experienced electricians will not use a knife to strip the sheath off; they will use a pair of side cutters and pliers. For this method the cable is first nicked in the end with the side cutters. The cpc is then pulled up through the slot with pliers.

By continually pulling the cpc back it will cut through the sheath and release the other cores (Figure 3.5).

There is nothing wrong with this providing care is taken not to damage the cpc.

Figure 3.5   *cpc being used to strip the sheath from the cable*

Remember
DO NOT
- damage the insulation
- damage the conductor
- cut the sheath back too far
- cut the insulation too deep so that you damage the conductor

---

The insulation is best removed using cable strippers (Figure 3.6). These must first be adjusted so that the blade only cuts into the cable insulation and not the conductor.

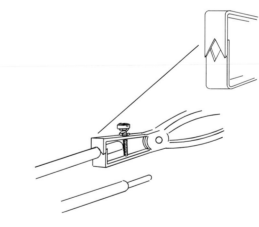

Figure 3.6

The cutting blades are now placed over the core and squeezed so that the blades cut into the insulation but stop before they reach the conductor. The insulation is now pulled off the conductor.

The insulation can be removed by a knife but care must be taken to avoid cutting your hand, damaging the conductor, or damaging the other cores.

Terminating the cable

When sheathed cable is being terminated the sheath must enter the enclosure and no exposed cores must be left showing. As the circuit protective conductor is not normally insulated a green and yellow sleeving must be slid over it so that it is insulated throughout its length within an accessory.

There are a number of different types of connections for the conductors (Figures 3.7–3.10).

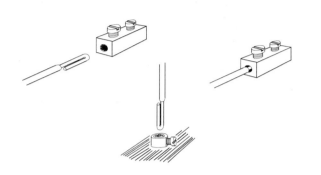

Figure 3.7   Tunnel connections

Figure 3.8   Wrap round connections

Figure 3.9   Crimp lugs

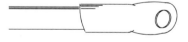

Figure 3.10   Solder lug

Entry into boxes and pattresses

Steel boxes should be fitted with a grommet (Figure 3.11) so that no sharp edges can damage the cable.

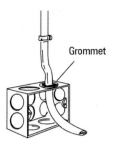

Figure 3.11   Cable entering a steel box

Plastic pattresses with "knockout" sections should have the section removed so that there are no sharp edges and the hole is the same size as the cable (Figure 3.12).

Figure 3.12   Cable entering a plastic pattress

This is best carried out using a junior hacksaw to determine the width of the hole and a pair of pliers to remove the plastic between the two cuts.

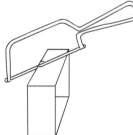

Figure 3.13   Cut the cable entry on a plastic pattress

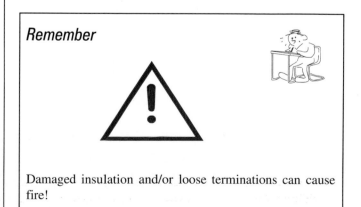

PVC insulated and sheathed flexible cable

Cable construction

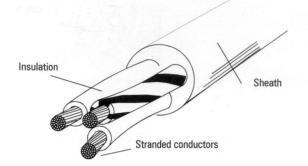

Figure 3.14

Stripping the cable

Care needs to be taken not to damage the insulation or the fine strands of conductor.

Using a solid knife "score" a ring around the circumference of the cable NOT going right through the cable sheath (Figure 3.15).

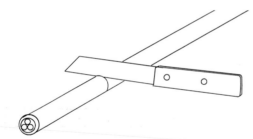

Figure 3.15   Do NOT cut right through – only score

Bending the cable on the score mark will now finish the separation of the sheath (Figure 3.16).

Figure 3.16   Bend to break the remaining PVC sheathing

The part that is to be removed can now be pulled over the cores and completely removed as a tube. This will only work for up to about 50 mm of sheathing at a time.

The insulation can be removed from the cores in the same way as for solid wiring cables.

Shaping and bending

There are set rules when it comes to shaping a cable for a particular environment.

Bends

If a bend is made too tight the conductors will be forced onto each other and the insulation is put under stress. Although it may appear to work without fault just after it has been completed, sometime in the future the insulation could break down and cause a serious hazard.

The bend should always lie flat against the surface and not fold or twist (Figure 3.17).

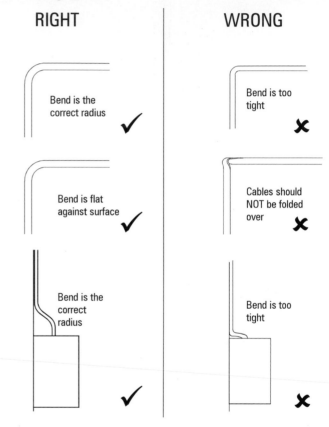

Figure 3.17

Clipping and fixing

Cables are clipped to a surface for several reasons:
- to support and reduce the stress on any point of the cable
- to keep the cable away from mechanical damage
- to make the cable run as neat as possible

Position of cable clips

The maximum distance between clips is determined by the cable's physical size across its widest side and whether the cable run is vertical or horizontal. For example a cable with a side dimension of 12 mm would have a maximum distance between clips horizontally of 300 mm and vertically 400 mm, whereas with a similar cable 18 mm wide the clips only need to be 350 mm for a horizontal run and 450 mm for a vertical run.

In most cases if the cable is clipped so that it is kept neat the maximum distances will automatically be achieved. To keep the appearance of a cable run acceptable there are several points that need to be considered:

- place clips equal distance either side of a bend NOT on the bend itself (Figure 3.18)
- where there is a short run work out the approximate position of the clips so that they can be fitted at equal distances before you start clipping (Figure 3.19)
- where cables enter accessories and a small set has to be made in the cable, keep clips back about 25 mm before the set starts (Figure 3.20)
- in general where cables enter accessories clips should be no closer than 50 mm from the accessory (Figure 3.20)

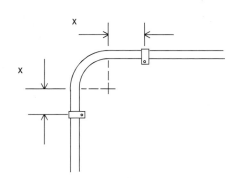

Figure 3.18   *Clips to be equal distances from the start and finish of the bend*

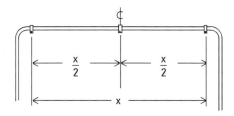

Figure 3.19   *Measure the distance between the clips at each end, then work out the number of clips required and make sure that they are equally spaced.*

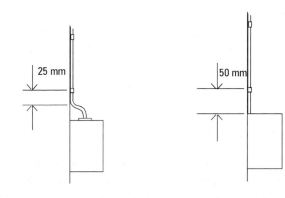

Figure 3.20   *Cables entering accessories*

Circuits

Where twin and cpc cable is used, or three core and cpc, circuits should be used that keep the number of cable connections to a minimum. The most common circuit used is the 3 plate (Figures 3.21 and 3.22).

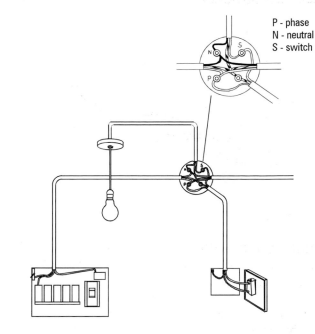

P - phase
N - neutral
S - switch

Figure 3.21

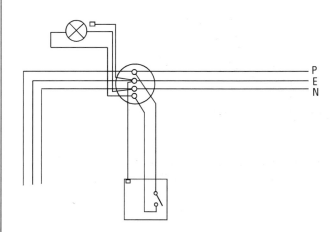

P
E
N

Figure 3.22   *3 plate lighting circuit diagram*

Note: Where core conductors other than red are used as switch wires they should be correctly colour coded.

13
Connecting a 13 A plug

Objective: To strip a flexible cable and connect a 13 A plug to an acceptable standard.

Suggested time: 15 minutes

Note: L (phase) brown, N (neutral) blue, E (earth) green/yellow

Materials:
1 short length of 3 core PVC sheathed flexible cable
A 13 A plug

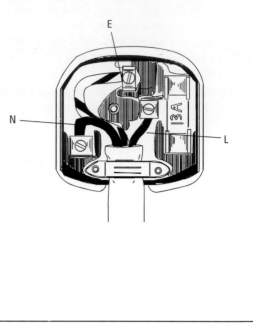

Points to consider:
- have the conductors been damaged during the stripping operation?
- has the insulation been damaged during the stripping operation?
- does the insulation go up to the terminals?
- are each of the conductors the correct length?
- are there any stray conductors that may be dangerous?
- does the sheath go into the cord grip?
- if the cable is pulled does it pull out of the plug in any way?
- are the conductors electrically and mechanically sound?

14
Connection of a junction box

Objective: To strip cables and connect up a junction box to an acceptable standard.

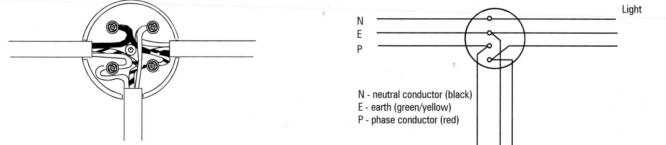

N - neutral conductor (black)
E - earth (green/yellow)
P - phase conductor (red)

Suggested time: 30 minutes

Materials:
3 short lengths of 1.0 mm^2 PVC insulated and sheathed 2 core and cpc cable
1 four terminal junction box

Points to consider:
- is the cable sheath of each cable inside the junction box?
- is any insulation damaged?
- does the insulation of each conductor extend up to the terminal?
- are the circuit protective conductors sheathed with green/yellow sleeving?
- is any conductor protruding through the terminal?
- is there enough "slack" on each conductor to prevent mechanical strain?
- are all connections tight?

15
Connection of a 13 A spur outlet unit and flexible cable

Objective: To connect a 13 A spur outlet unit to two 2.5 mm^2 cables and a 1.5 mm^2 flexible cable to an acceptable standard.

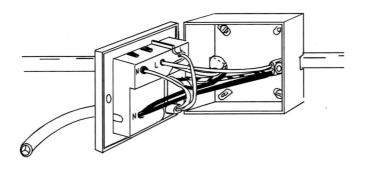

Suggested time: 45 minutes

Materials:
2 short lengths of 2.5 mm^2 PVC insulated and sheathed twin and cpc cable
1 short length of PVC insulated and sheathed three core flexible cable
1 × 13 A spur outlet unit

Points to consider:
- does the cable sheath in each case enter the unit?
- is any insulation damaged?
- does the insulation in each case go up to the terminal?
- is each circuit protective conductor sleeved?
- are the conductors cut and stripped to the correct length?
- are all connections tight?
- is the cord grip tight on the sheath of the flexible cable?

Two-way lighting circuit

Often it is necessary to have two switches controlling the same light. This system is usually found on staircases where the light can be switched on or off at the top or the bottom of the stairs.

There is more than one way of achieving this – the one shown in Figures 3.23 and 3.24 uses a 3 core and cpc cable to convert an existing one-way system to a two-way. A two-way switch has three terminals, one is labelled the common. This must always be treated separately.

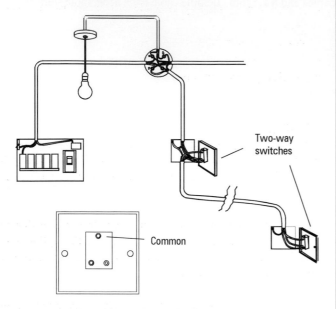

Figure 3.23   *Two-way switch showing connections*

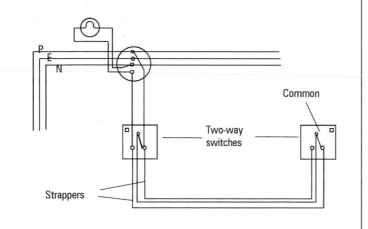

Figure 3.24   *Two-way lighting circuit diagram*

Intermediate switching circuit

Within some installations it is not enough to have only two switches controlling one light. Where there are a number of doors into one room or there is a staircase with rooms off of a landing part of the way up, then a third or even fourth switch may be required. Although there is more than one way of doing this the basic theory is the same.

An intermediate switch is connected into the strappers of a two-way switched system (Figure 3.25).

The switch has four terminals and the internal switch connections are as shown in Figure 3.26.

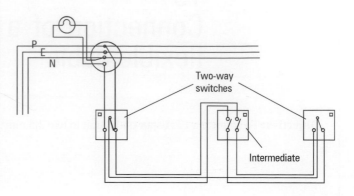

Figure 3.25   *Two-way and intermediate switched circuit diagram*

| Switch connections | Switch position | Switch connections | Switch position |

Figure 3.26

The ring final circuit

As the name implies, the circuit starts from and finishes at the same terminals and should form a continuous ring. The phase starts and finishes at the same protective device (fuse or circuit breaker), the neutral starts and finishes at the same terminal on the neutral bar, and the circuit protection conductor starts and finishes at the same connection on the earth block. This type of circuit is generally used for socket outlets and so each socket outlet on the ring has at least two cables to it. This means that there must be at least two conductors in each terminal. The phase conductors being connected to the L terminal, the neutral to the N terminal and the circuit protective conductors to the E terminal. Where a socket is required at a point away from the natural run of the ring circuit cables a non-fused spur may be used. This is a single cable run just to this outlet. The cable conductors must not be of a smaller cross sectional area than that of the ring conductors. Usually in domestic premises a 2.5 mm^2 conductor is used with a 1.5 mm^2 circuit protective conductor. The non-fused spur may be connected from:

i)   a socket on the ring

ii)   a junction box connected onto the ring or

iii)   the consumer unit or distribution board

In general there should be a separate ring final circuit for every 100 m^2 of floor area. It must not be forgotten that the total load that can be connected to a ring protected by a 30 A or 32 A fuse is just over 7 kW.

A non fused spur must not supply more than one single **OR** one twin socket outlet (Figure 3.27).

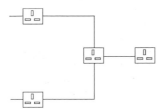

Non-fused spur from a 13 A socket outlet

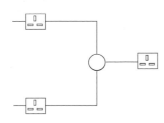

Non-fused spur from a junction box

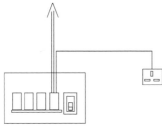

Non-fused spur from a distribution board

Figure 3.27   *Non-fused spurs*

Non-fused spurs are wired in cable of the same rating as the ring circuit, but where fused spurs are connected, the rating of the cable may be reduced accordingly.

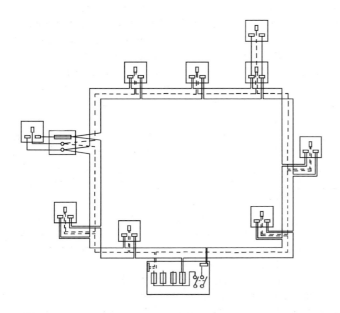

Figure 3.28

Inspecting and testing

Inspection

All installations need inspections during and on completion of work. The items to be considered on circuits include the:

- connection of conductors
- correct identification of conductors
- correct cross sectional area of conductors
- connection of single-pole devices in the phase conductor only
- correct connection of accessories
- correct overcurrent device type and rating
- labelling of the circuits

Testing

An inspection is the initial check to ensure the installation is correctly assembled. Testing must be carried out to verify the installation meets the requirements of the design and it is safe to energise.

The first test that must be carried out is to confirm that the circuit protective conductors are continuous throughout each circuit. The test equipment required for this is an ohmmeter with a low resistance scale.

This test is confirming that the earth connection in the consumer unit is directly linked to all earth connections throughout the circuit and to any exposed metalwork associated with it. Circuits in the exercises included in this section use pendant type lamp connections but often metal cased fluorescents are used. In these circumstances the metal enclosure should be connected to earth and checks made to ensure that it is.

Continuity of protective conductor test

There is more than one way to carry out the continuity of protective conductor test and it will depend on the particular circumstances as to which will be the most suitable.

Method 1, the $R_1 + R_2$ method, is useful to assess the continuity of circuit protective conductors, correct polarity of the circuit and establish the value of $R_1 + R_2$, which may be required to determine earth fault loop impedance. Method 2, the long lead method, is more appropriate for testing the continuity of main and supplementary bonding conductors. For both methods the supply must be securely isolated and any relevant documentation must be available before starting the test.

The test equipment required is a continuity test instrument with a low resistance scale, capable of measuring values in milliohms.

Re*member*

R_1 is the resistance of the phase conductor and R_2 is the resistance of the cpc.

Test method 1

After securely isolating the supply disconnect any supplementary bonding which may affect the readings. Make a temporary link (Figure 3.29) at the supply end between the phase and the cpc of the circuit under test. Now measure the resistance between the phase and cpc at each point, such as light fittings, switches and so on, in the circuit. The value obtained is known as the $R_1 + R_2$ value and the highest measured value for each circuit should be recorded.

After completing the test remove the temporary link at the supply end of the circuit and reconnect any bonding conductors as necessary.

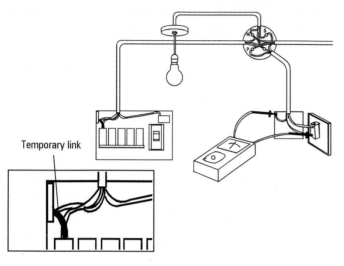
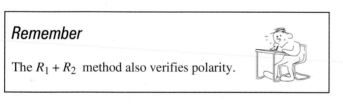

Temporary link

Figure 3.29   *Test method 1 ($R_1 + R_2$)*

Remember

The $R_1 + R_2$ method also verifies polarity.

Test method 2

Securely isolate the supply. From the consumer's main earth terminal attach a long test lead to the ohmmeter. With the other ohmmeter lead make contact with the protective conductors at points on the circuit (Figure 3.30). In this method, as the distance between testing points can be quite long, the resistance of the test leads must be taken and this reading subtracted from each test result. For example if a test resistance is 0.7 Ω and the test leads have a resistance of 0.2 Ω then the actual resistance of the test is 0.7 – 0.2 = 0.5 Ω.

Alternatively, some meters can be zeroed with the long leads connected.

Record the values taken and deduct the values of the test leads as above. The resistance of the protective conductor R_2 should be recorded on the appropriate form.

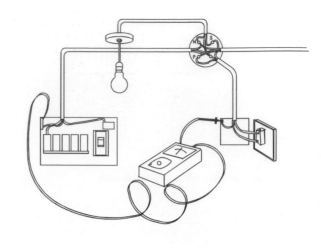

Figure 3.30   *Test method 2 (long lead)*

Continuity of ring final circuit conductors

The test for continuity of ring final circuit conductors is to verify that the ring is complete and has not been interconnected. If they are not connected correctly ring final circuits could become a fire hazard.

There are several methods of carrying out this test and at this stage we will only be looking at the one detailed in IEE Guidance Note 3, Inspection and Testing.

The tests are carried out before the supply is connected using a low reading ohmmeter. The first test is to confirm that there is a circuit between the two ends of the ring final circuit cables. In this circuit there are three separate rings, one for the phase, one for the neutral and one for the circuit protective conductors. If the protective circuit had been in steel conduit or trunking this would not have to be connected in a ring.

To verify the continuity of the ring final circuits they are first disconnected at the distribution board and brought out in their pairs. The resistance of the conductors are taken, and recorded, phase to phase, neutral to neutral and cpc to cpc. This measures the resistance, end to end, of each conductor. The values for the phase and neutral conductors should be about the same value. Generally the cpc is a reduced cross-sectional area (for example 2.5 mm^2 phase and neutral with 1.5 mm^2 cpc) and so the value for the cpc should be higher, around 1.7 times that of the phase. This test determines that the circuit is a ring and the correct conductors have been identified. If appropriate values are not obtained further investigation is required.

The phase conductor of one pair should then be connected to the neutral of the other and the remaining phase and neutral conductors should be similarly connected. The resistance is measured between the cross connected pairs (Figure 3.31).

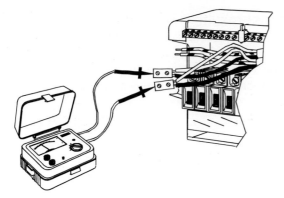

Figure 3.31   Ring circuit continuity test

The reading obtained should be approximately half the values we obtained measuring phase to phase or neutral to neutral. With the phase and neutral conductors still "cross connected" a reading should be taken at each socket on the ring. The values obtained should be substantially the same as those we had at the distribution board. Remember that the resistance at outlets spurred from the ring will be higher, dependent upon the length of cable in the spur.

We now cross connect the phase and cpc conductors in the same way and take a reading at the distribution board (Figure 3.32). This value should be approximately

$$\frac{\text{phase to phase } \Omega + \text{cpc to cpc } \Omega}{4}$$

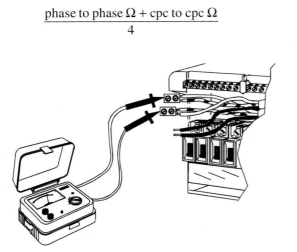

Figure 3.32   Ring circuit continuity test

A reading is then taken at each socket outlet, which, as before should be the same as that obtained at the distribution board.

This will ensure that the conductors are connected as a ring circuit and confirm polarity at each socket.

Insulation resistance

The first two tests we have looked at were to prove that there was a circuit and that the resistance was low enough.

The insulation resistance test (Figure 3.33) is to confirm that the insulation throughout the circuit has not been damaged.

Before the test is carried out it is important to check that
• there is no supply to the circuit being tested
• all lamps are removed
• all equipment that would normally be in use is disconnected
• any electronic equipment that would be damaged by the high voltage test is disconnected (this may include lamp dimmer switches, delay timers, and so on)
• all fuses are in place
• all switches are in the ON position (unless they protect equipment that cannot otherwise be disconnected)

For installations supplied at voltages up to 500 V a.c. the test voltage is 500 V d.c.

The test must be carried out between:
• phase and neutral
• phase and earth
• neutral and earth
• phase and phase of a three-phase circuit

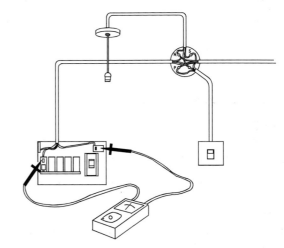

Figure 3.33   Insulation resistance test

Each reading should be at least 0.5 MΩ, on a circuit such as those included in this section, to comply with BS 7671. However where an insulation resistance is recorded at less than 2 megohm there is a possibility of a latent defect. The reading may indicate INF or ∞ (infinity). This means the reading is greater than the instrument can indicate. Values should not be recorded as ∞ or INF but as "greater than" the highest reading on the instrument scale. So if, for instance, the instrument maximum reading is 50 MΩ and the result is higher than that at ∞, we should record > 50 MΩ, where ">" indicates "greater than".

Polarity

A polarity test (Figure 3.34) checks that the installation connections have been made in the correct conductors. Much of the polarity test can be done when doing the $R_1 + R_2$ test method for the continuity of protective conductors.

On domestic installations the tests should verify that:

- all lighting switches are connected in the phase conductor
- the centre pin of any ES (Edison Screw) lampholders are connected in the phase conductor
- the switches on socket outlets are connected in the phase conductor
- the correct pin of socket outlets is connected to the phase conductor
- where double pole switches are used, such as for immersion heaters, the phase and neutrals have not been swapped over.

A visual inspection should have confirmed the correct identification of the conductors so this should be assumed correct unless proved otherwise. This can be carried out with a continuity test instrument or a bell set.

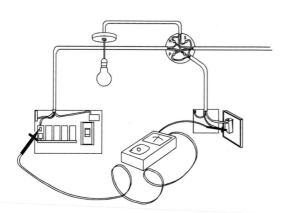

Figure 3.34   Polarity test

Earth fault loop impedance

This test should be carried out on all circuits. It is very important that, if a fault does develop between phase and earth, the protection device operates in the shortest possible time. Where portable hand held equipment is in use this time must not exceed 0.4 seconds. In other circuits it can be as long as 5 seconds. Some special installations require disconnection to be within 0.2 seconds.

On TN-systems the injected current flows through the consumer's circuit protective conductors and main earth. It then passes through the supply authority's earth back to the centre point of their transformer. The return path of the current is through the phase conductor back to the test equipment.

This path is the same as that of phase to earth fault currents, so if the resistance of this is kept very low faults should clear within the required time.

The test results

When the inspection and testing have been completed, the results must be recorded on the form prescribed in BS 7671 Appendix 6 and, together with a schedule of test results, forwarded to the person ordering the report.

16
Wiring a 3 plate lighting circuit

Objective: To wire, connect and test a 3 plate lighting circuit.

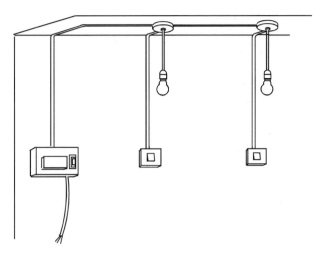

Suggested time: 4 hours

Materials:
1 consumer unit and a 5 or 6 A protection device and blanks as required
2 × 3-plate ceiling roses
2 one-way switches with surface mounting boxes
2 lamps
green/yellow sleeving
1 measured length 2-core heat resisting 0.5 mm^2 flexible cable
2 lamp holders
1 measured length 1.0 mm^2 PVC insulated cable
cable clips
fixings
red sleeving

Points to consider:
- is any of the cable insulation damaged?
- do all of the cable sheaths enter the enclosures?
- are all conductors left at an acceptable length?
- are all of the clips placed at acceptable distances?
- are all circuit protective conductors sleeved throughout their length?
- are all terminations tight?
- does the circuit work correctly?
- is the polarity correct?
- are the correct fuseway and neutral bar terminal used in the consumer unit?
- are flexible cable grips used at the ceiling rose and lampholder?
- are the conductors doubled over at the terminations?

17
Wiring a two-way and intermediate lighting circuit

Objective: To convert a 3 plate single way lighting system into a two-way and intermediate system.

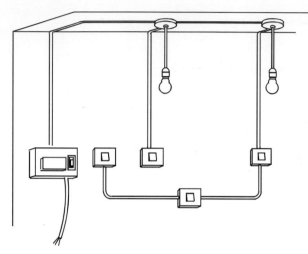

Suggested time: 2 hours – after the completion of the previous exercise

Materials:
1 consumer unit* (completed)
2 × 3-plate ceiling roses*
2 lamp holders*
1 measured length 2-core flex*
1 measured length twin 1.0 mm^2 cable*
cable clips*
green/yellow sleeving
1 measured length 3-core and cpc 1.0 mm^2 cable
2 two-way switches and 1 intermediate switch with surface mounting boxes (× 2)
fixings
red sleeving

* already installed in the previous exercise

Points to consider:
- is any of the cable insulation damaged?
- do all of the cable sheaths enter the enclosures?
- are all conductors left at an acceptable length?
- are all of the clips placed at acceptable distances?
- are all conductors at the correct length?
- are all circuit protective conductors sleeved throughout their length?
- are all terminations tight?
- is the circuit correct when tested with a meter?
- does the circuit work?
- is the polarity correct?
- are the correct fuseway and neutral bar terminal used in the consumer unit?
- are flexible cable grips used at the ceiling rose and lampholder?
- are the conductors doubled over at the terminations?

18
Wiring a ring final circuit

Objective: To wire a ring final circuit as shown.

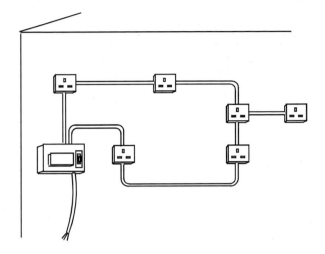

Suggested time: $2\frac{1}{2}$ hours

Materials:
1 consumer unit and 30 A or 32 A protective device and blanks as required
6 socket outlets with surface mounting boxes
green/yellow sleeving
a measured length 2.5 mm^2 PVC twin and cpc cable
cable clips
fixings

Points to consider:
- is any of the cable insulation damaged?
- are all of the cable clips placed at acceptable distances?
- are all of the protective conductors sleeved throughout their length with green/yellow sleeving?
- are all terminations tight?
- is the circuit correct when tested with a meter?
- does the circuit work?
- is the polarity correct?
- are the correct fuseway and neutral bar terminal used in the consumer unit?
- is there a suitable amount of spare cable left in the consumer unit?
- is there a suitable amount of spare cable left at the accessories?

19
Earthing and bonding

Objective: To carry out bonding between a consumer's earth terminal, the consumer's main control equipment, main gas intake and main water intake.

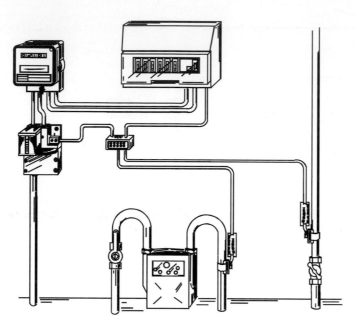

Suggested time: 45 minutes

Equipment:
Main electrical, water and gas equipment

Materials:
2 bonding clamps
1 measured length of 4 mm^2 yellow/green single cable
1 measured length of 6 mm^2 yellow/green single cable
1 measured length of 10 mm^2 yellow/green single cable
1 measured length of 16 mm^2 yellow/green single cable
cable clips

Points to be considered:
- has the conductor with the correct cross-sectional area been used?
- are all sections bonded?
- are all connections tight?
- have the gas and water pipes bonding connections been made on the correct side of their main control?
- are the clamps in an acceptable position?
- have the bonding leads been clipped to an acceptable standard?
- have the connections been labelled in accordance with specific requirements?
- has any paint been removed and have the pipes been cleaned?

20
Visual inspection

Objective: To carry out visual inspections to identify good and bad installation practice.

Inspections:

All installations need constant inspection during and on completion of work. The items to be checked include the:

- connection of conductors
- correct identification of conductors
- correct cross-sectional area of conductors
- connection of single-pole switches in the phase conductor only
- correct connection of lampholders
- correct overcurrent device type and rating
- labelling of the circuits

The inspection should be carried out on PVC sheathed cables at accessories and throughout the installation where possible. Guidance note 3 provides some advice on inspecting sampling for larger installations.

Points to be considered:
- is the insulation the correct length?
- are all connections tight?
- are the protective conductors correctly sleeved?
- are the conductors correctly identified?
- are there exposed conductors showing?
- did the inspection conform to regulatory requirements, codes of practice, guidelines?
- were all the specified inspections completed?

21
Continuity of protective conductors test

Objective: To carry out tests to measure the continuity of protective conductors.

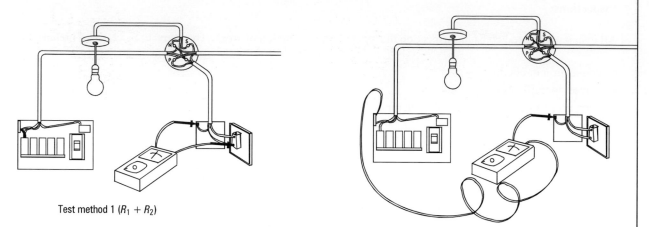

Test method 1 ($R_1 + R_2$)

Test method 2 (long lead)

Suggested time: dependent upon the installation or circuit being tested.

Method:
The method chosen should be the most appropriate for the particular circumstances either Test method 1 or Test method 2 as detailed on page 48.

Equipment:
a selection of test instruments both suitable and unsuitable for this test.

Points to be considered:
- has the correct instrument been selected?
- has the serial number of the selected instrument been noted?
- are the batteries in the selected test equipment charged to an acceptable level?
- should the instrument be zeroed?
- has the instrument been tested by short circuiting the leads before use?
- has the resistance of the test leads been noted?
- are the resistances of the circuit protective conductors acceptable?
- has the temporary link been removed on completion of Test method 1?

22
Ring final circuit continuity test

Objective: To carry out a ring final circuit continuity test using suitable test equipment.

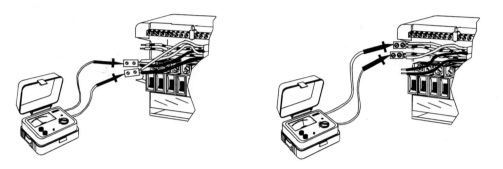

Suggested time: dependent upon the installation being tested.

Method:
Carry out test of ring final circuit continuity (as detailed on page 46).

Equipment:
A selection of test instruments both suitable and unsuitable for this test.

Points to be considered:
- has the correct instrument been selected?
- has the serial number of the selected instrument been noted?
- are the batteries in the selected test equipment charged to an acceptable level?
- should the instrument be zeroed?
- are the test leads in good condition?
- is the circuit isolated from the supply?
- are the values of resistance such that the ring final circuit is proved?

23
Insulation resistance test

Objective: To carry out an insulation resistance test with suitable test equipment.

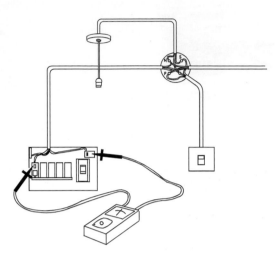

Suggested time: dependent upon the installation being tested.

Method:
Carry out an insulation resistance test (as detailed on page 49).

Equipment:
A selection of test instruments both suitable and unsuitable for this test.

Points to be considered:
- has the correct instrument been selected?
- has the serial number of the selected instrument been noted?
- are the batteries in the selected test equipment to an acceptable level?
- should the instrument be zeroed?
- are the test leads in good condition?
- has the instrument been tested open circuit and short circuit before use?
- is the circuit isolated from the supply?
- is there any danger to any other person if the test is carried out?
- were all circuit switches in the ON position before testing?
- are the readings acceptable?

4

Metal Sheathed Cable Systems

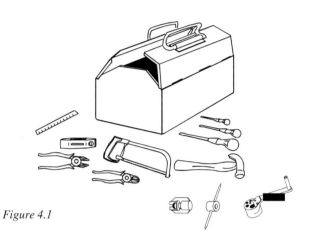

Figure 4.1

On completion of this chapter you should be able to:

◆ strip mineral insulated cable to an acceptable standard
◆ terminate a mineral insulated cable
◆ test a mineral insulated cable for insulation resistance and polarity
◆ install mineral insulated cable to specified dimensions
◆ wire an "open circuit" alarm circuit
◆ strip PVC steel wire armoured cable
◆ terminate a steel wire armoured cable

Part 1
Mineral insulated cable

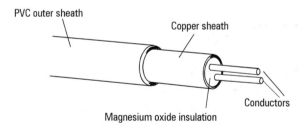

Figure 4.2   *Cable construction*

The tool kit required

For stripping the cable
• junior hacksaw
• pliers
• side cutters
• MI stripping tool

For terminating cable
• MI pot wrench
• MI crimping tool
• gland pliers
• spanners

For fixing cable
• screw driver
• hammer
• wood block

For testing
• 500 V insulation resistance test instrument
• continuity test instrument

Stripping the cable

The outer PVC sheath is not included on all cables but if it is this must first be cut back to give enough exposed cable to allow for termination. It is essential to terminate this type of cable in a dry atmosphere as the insulation will absorb moisture.

To expose the conductor the copper sheath must be removed. There are a number of special "stripping" tools available for this purpose (Figure 4.3). Each type of tool works on a principle similar to a tin opener. A blade cuts into the copper and peels it off as the stripping tool is rotated.

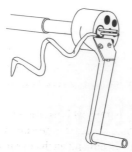

Figure 4.3   *Cable stripping tool*

When the correct length of stripped cable has been reached a pair of pliers can be used to terminate the process and leave a clean cut on the cable sheath (Figure 4.4).

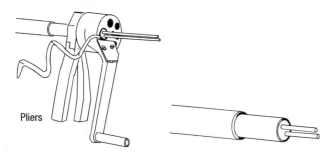

Pliers

Figure 4.4

Terminating the cable

A complete termination and seal are shown in Figure 4.5.

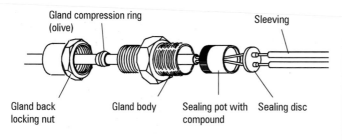

Gland compression ring (olive)
Sleeving
Gland back locking nut
Gland body
Sealing pot with compound
Sealing disc

Figure 4.5

If a complete termination is being used, then the termination gland should be slid onto the cable in the correct order.

The sealing pot (Figure 4.6) should be screwed onto the sheath of the stripped cable. As the gland and pot are sized for the particular csa and no. of cores in each cable the correct size must be used for the cable being terminated.

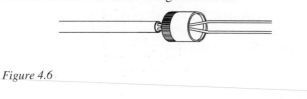

Figure 4.6

A pot wrench (Figure 4.7) may be used for this purpose.

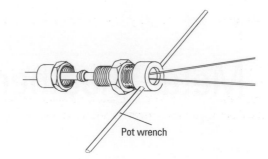

Pot wrench

Figure 4.7

The sealing pot must now be filled with sealing compound (Figure 4.8). Care must be taken at this stage to eliminate any dirt or air pockets inside the sealing pot.

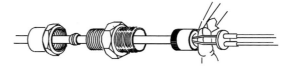

Figure 4.8

This must now be sealed with the plastic disc and crimped so that it cannot be pulled off (Figure 4.9).

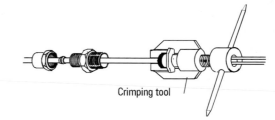

Crimping tool

Figure 4.9

The conductors now require an insulated sleeving for their full length up to where they will be finally connected to a terminal (Figure 4.10).

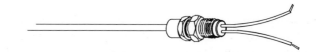

Figure 4.10

As the conductors are not marked as to which is phase and neutral it is necessary to do this. However before this can be done tests must be carried out to ensure that there are no connections between the cable conductors or between conductors and cable sheath.

Testing

This test must be carried out at at least 500 V d.c. on a 230 V supply.

First each end of the cable must be checked to make sure the conductors are insulated from each other throughout the cable and terminations. The 500 V insulation resistance tester is now connected between the conductors and a reading taken (Figure 4.11).

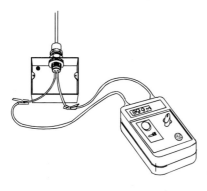

Figure 4.11

This reading must not be less than 0.5 MΩ. On a new cable the value would be expected to be considerably higher, in the region of 100 MΩ. Where there are more than two conductors tests will need to be carried out between all of them. When it has been established that there is no circuit between conductors a test must check that no conductor is in contact with the metal sheath. The quickest way to do this is to connect all the conductors together and test between them and the copper sheath (Figure 4.12).

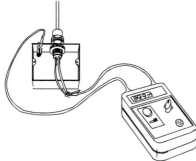

Figure 4.12

The result of this test must be greater than 0.5 MΩ. If there is a low reading then it will be necessary to disconnect the conductors and test between each one and the cable sheath separately.

In the case of a serious fault the sealing pot may have to be removed from the cable and an examination of the seal will need to be carried out to find the cause of the breakdown.

Polarity test

Unlike PVC sheathed cables the cores are not identified throughout their length. This means that they must be tested out and labelled. The length of some cable runs makes it difficult to connect meter leads to each end of the cable.

The best way to carry out the test is to connect one of the cables to the metal box or cable sheath at one end of the cable. Then test between the metal sheath and each core in turn at the other end, as shown in Figure 4.13. When the core has been identified it should be marked with coloured sleeving, number tag or phase tape.

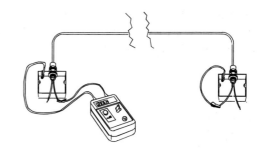

Figure 4.13

> **Remember**
> **It is essential to terminate mineral insulated cable in a dry atmosphere, as the cable is hygroscopic and so will absorb moisture.**

Connection to accessories

In standard accessory boxes where there is a 20 mm hole, the complete termination can be used and tightened with locknuts (Figure 4.14) or a coupler and male bush (Figure 4.15).

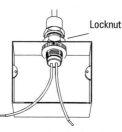

Locknut

Figure 4.14

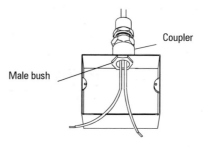

Coupler

Male bush

Figure 4.15

The method used may depend on the size of the accessory using the box (Figure 4.16).

Figure 4.16   Termination into a spouted conduit box

Bends

The radius of bend on mineral insulated cable is critical, for a bend made too sharp may damage the cable. The radius of the bend should never be less than 6 times the overall diameter of the cable (Figures 4.17 and 4.18).

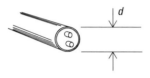

Figure 4.17   Overall diameter of cable

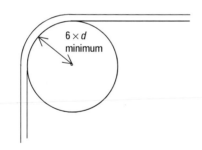

Figure 4.18   Radius of the bend NEVER less than 6 times the overall diameter of the cable

The firmness of the cable can be very useful when fixing it to the surface. The cable does however get bent in the wrong places and then has to be straightened out. On long lengths where this has to be done there are hand operated machines which run the cable through rollers and take out unwanted bends and kinks. On short runs these unsightly sections can be straightened by hand or by the use of a hammer and a block of wood (Figure 4.19).

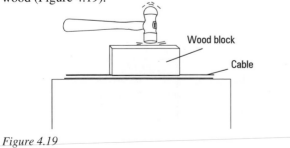

Figure 4.19

The cable should never be hit directly by the hammer (Figure 4.20) as this could damage and split the metal sheath.

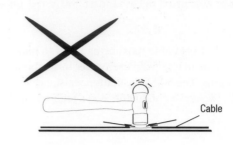

Figure 4.20   NEVER hit the cable directly with a hammer

Cable supports

The rules used for fixing PVC cables apply to mineral insulated cables but as these are more rigid not as many clips are necessarily needed. The clips used are either saddles or "P" clips (Figures 4.21 and 4.22).

Figure 4.21   Saddle

Figure 4.22   "P" type clip

Alarm circuits

Mineral insulated cable is ideal for fire alarm circuits as it will carry on working in very high temperatures. The basic circuit used on many fire alarms is the same as security circuits or other warning circuits.

Fire alarm detectors

In many situations humans detect fires before automatic devices operate. In these cases "break glass" manual devices (Figure 4.23) set off the alarm.

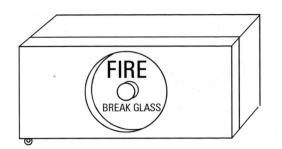

Figure 4.23

In order to protect life and property when constant human presence is not possible there are many automatic fire detectors available including:

- heat detectors
- smoke detectors
- flame detectors

There are two basic types of alarm circuits – "open circuit" and "closed circuit". However, more sophisticated circuits are often used for fire and security installations.

Open circuit

This is the basic circuit used for a front door bell (Figure 4.24).

When the push is operated the bell rings. It stops when the push is released. This same circuit could be a fire alarm where the push is a "break glass" contact or smoke detector, and the "push" is held on until it is reset.

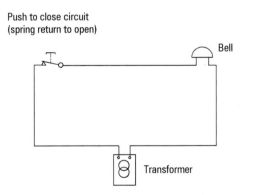

Figure 4.24

Closed circuit

On this type of circuit the push or detector is closed until it is activated, and because of this it has to be independent of the bell or sounder circuit. The push circuit and the bell circuit are linked through a relay.

Two separate supplies are shown in Figure 4.25, however they could be taken from the same source.

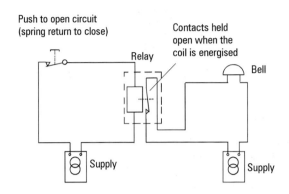

Figure 4.25

24
MIMS cable terminations

Objective: To make off one complete termination and one seal only to a measured length of MIMS cable and test.

Suggested time: 1¼ hours

Materials:
1 complete MIMS termination
1 seal assembly for MIMS cable
1 measured length of 2 core 1 mm² MIMS cable

Points to consider:

- was the cable stripped without damage?
- were the seals correctly terminated?
- was the termination correctly assembled?
- was an acceptable length of conductor left?
- did the insulating sleeve cover the conductors?
- was the insulation resistance value acceptable?
- was the polarity of the cable correctly identified?

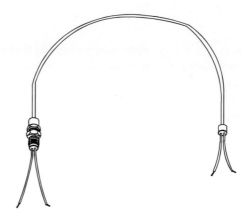

NOTE: This cable will be used as part of a later assessment. The length of the tails is dependent on the accessory used.

25
Terminating and fitting MIMS cable

Objective: To terminate both ends of a measured length of MIMS cable as shown, test and identify conductors.

Suggested time: 2 hours

Materials:
2 complete MIMS terminations
1 metal switch box
1 steel conduit 4 spout cross box
1 measured length of 2 core 1 mm² MIMS cable
cable clips
500 V insulation resistance test instrument

Points to consider:

- were the seals correctly terminated?
- was the gland assembled correctly?
- was an acceptable length of conductor left?
- was the cable bent to an acceptable radius?
- was the cable insulation tested correctly?
- were the clips spaced at suitable intervals?
- was the insulation resistance value acceptable?
- were the conductors correctly identified?
- were the conductors correctly identified?

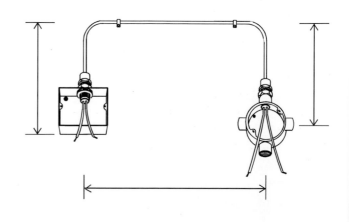

NOTE: This cable will be used as part of a later assessment.

26
Inspection and testing, the polarity test

Objective: To carry out a polarity test with suitable test equipment.

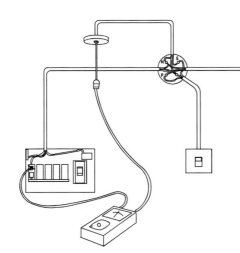

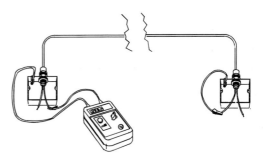

Method – PVC/PVC cables

It is important to have all single pole devices connected in the phase conductor only. This test is to confirm this. A visual inspection should have confirmed the correct identification of the conductors so this should be assumed correct unless proved otherwise. This can be carried out with a continuity test instrument.

Method – MIMS cable systems

Unlike PVC sheathed cables the cores of MIMS cables are not identified throughout their length. This means that the cores must be tested and labelled. The length of some cable runs makes it difficult to connect meter leads to each end of the cable.

The best way to carry out the test is to connect one of the cables to the metal box or cable sheath at one end of the cable. Then test between the metal sheath and each core in turn at the other end. When the core has been identified it should be marked with coloured sleeving, number or phase tape.

Suggested time: related to the installation being tested.

This can be carried out on all Exercises that include circuit wiring.

Equipment:

A selection of test equipment both suitable and unsuitable for this test.

Points to consider:

- has the correct equipment been selected?
- has the serial number of the selected instrument been noted?
- were the batteries (if included) to an acceptable level?
- was the polarity correctly identified?
- did the circuit work as it should?

27
Alarm circuits in MIMS cable

Objective: To use the cables terminated in previous exercises to wire two alarm circuits, one closed and one open circuit alarm, and test each circuit for correct wiring.

Suggested time: $2\frac{1}{2}$ hours

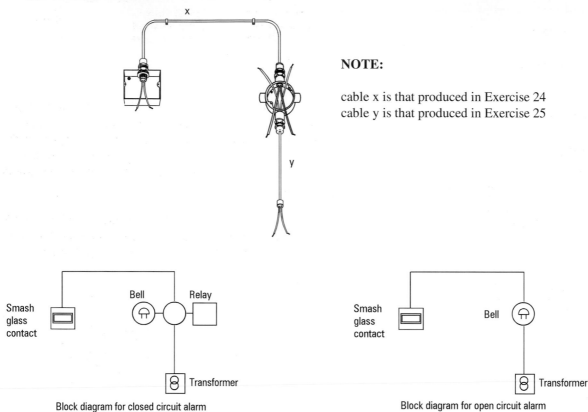

NOTE:

cable x is that produced in Exercise 24
cable y is that produced in Exercise 25

Block diagram for closed circuit alarm

Block diagram for open circuit alarm

Materials:
(other than that used in Exercises 24 and 25)

For the closed circuit alarm exercise:
1 break glass contact (closed circuit type)
1 ELV bell
1 bell transformer with output suitable to drive the bell
1 "open when energised" relay
interconnecting single core cables

In addition for the open circuit alarm exercise:
1 break glass contact (open circuit type)

Points to consider:

- were the terminal connections made off correctly?
- was the circuit connected correctly?
- did the circuit work when tested?

Part 2
Steel wire armoured cable (SWA)

The tool kit required

For stripping the cable

- hacksaw
- junior hacksaw
- knife (not Stanley type)
- pliers

in addition, for terminating the cable

- appropriate spanners
- gland pliers

in addition, for fixing the cable

- screw drivers

in addition, for testing

- test instruments

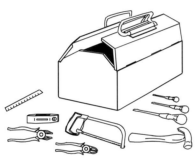

Figure 4.26

Cable construction

The steel armouring (Figure 4.27) on this cable protects the live conductors from mechanical damage and may also be used as a circuit protective conductor. In any event they form part of the electrical installation and need to be connected to the main earthing terminal, generally via the circuit protective conductors (cpc). As this type of cable is very rigid the termination often has to be made within a particular tolerance slack cable cannot be pushed back and there is no extra to spare. This means that the cable must be cut, stripped and terminated with some accuracy.

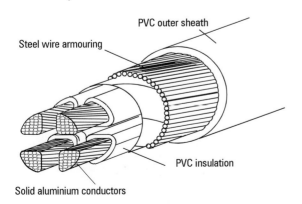

Figure 4.27

Steel wire armouring
PVC outer sheath
PVC insulation
Solid aluminium conductors

Stripping and terminating the cable

There are several different types of SWA termination but they all have some similar features (Figure 4.28).

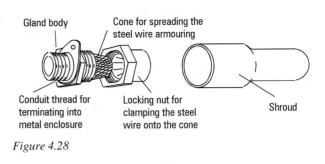

Gland body
Cone for spreading the steel wire armouring
Conduit thread for terminating into metal enclosure
Locking nut for clamping the steel wire onto the cone
Shroud

Figure 4.28

To terminate an SWA cable within a set distance the gland body needs first to be fitted into the enclosure in the position it will be when the termination is complete. The gland locking nut, and the shroud if one is used, need to be slid over the cable at this stage. These can be taped back out of the way for they will not be required yet.

Hold the cable where it will be installed and mark off the PVC sheath at a point above the base of the cone one thickness of the diameter of a steel wire as shown in Figure 4.29.

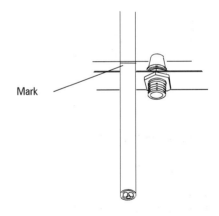

Mark

Figure 4.29

Cut a single line around the sheath at this point (Figure 4.30) using a junior hacksaw. Care should be taken to ensure that this cut is made only halfway through the steel wire armour.

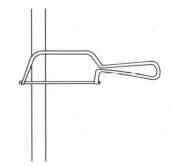

Figure 4.30

The PVC outer sheath from this cut line to the end of the cable is then removed. For smaller cables this may simply be pulled off, for larger cables the sheath will need to be stripped using a knife. With the outer sheath removed the armours can be removed at the cut. This is best achieved by bending the wires back and forth until they break. Care should be taken not to distort the steel wires within the sheath.

Once the wires have been cut offer the core against the remaining sheath as shown in Figure 4.31 and mark the sheath ready for removing the remaining section, this needs to be cut through and removed using a knife taking care not to damage the conductors or insulation.

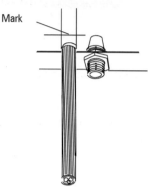

Mark

Figure 4.31

The cable now has to be put through the gland body into its completed position. The steel wire strands have to be spread out over the cone making sure none are crossing over others (Figure 4.32).

When everything is in place the gland locking nut can be put into place and tightened up.

The cable cores can now be removed from their PVC sleeve. Care must be taken not to damage the insulation around the conductors.

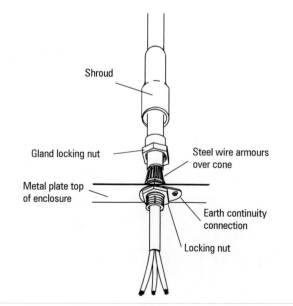

Shroud

Gland locking nut

Steel wire armours over cone

Metal plate top of enclosure

Earth continuity connection

Locking nut

Figure 4.32   PVC/SWA/PVC cable terminations

Bends

The construction of the cable limits the radius of bends which can be made in it. However, so that no damage occurs, minimum bending radii are laid down. These relate to the makeup of the conductor incorporated in the cable. If stranded cores are used the factor is 6 times the overall cable diameter (Figure 4.33). As solid core cables are not as flexible a factor of 8 times the overall diameter is used (Figure 4.34).

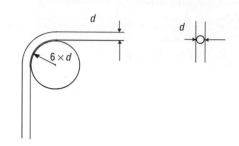

d

d

$6 \times d$

Figure 4.33   Stranded cables

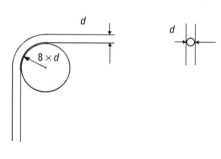

d

d

$8 \times d$

Figure 4.34   Solid core cables

Cable supports

The method of supporting SWA cables depends on the environment they are installed in. Where there are several cables run through the same area they are often fitted to a cable tray. If this is the case on horizontal runs the cable is supported throughout its length. It does, however, still require fixing to the cable tray.

On vertical runs the supports would have to be provided at the same distances regardless of the fact it is on cable tray. Examples of the horizontal and vertical maximum spacings of supports are shown in Table 4.1. The diameter of the cable is again important when determining the distances between supports. For flat cables the diameter is taken as the dimension of the major axis.

Table 4.1 Spacing of supports for cables in accessible positions Armoured cables

Reproduced from Table 4A, *IEE On-Site Guide*, by kind permission of the Institution of Electrical Engineers

Overall diameter of cable	Maximum spacing of clips	
	horizontal	vertical
mm	mm	mm
up to 9	—	—
over 9 up to 15	350	450
over 15 up to 20	400	550
over 20 up to 40	450	600

Many of the devices used for supporting SWA cables are the same as those used for conduit but the one designed specifically for SWA is the cable cleat as shown in Figures 4.35 and 4.36. This has a single fixing hole which may be used for wood screws or bolts.

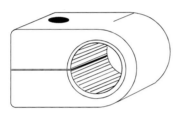

Figure 4.35

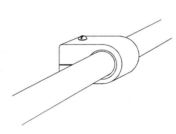

Figure 4.36

Aluminium foil sheathed cables
FP200 and PX

These cables consist of an outer sheath of PVC which is over a layer of aluminium foil (Figure 4.37). The cores of the cables are similar to other PVC sheathed and insulated cables consisting of insulated live conductors and a bare circuit protective conductor.

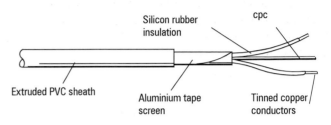

Figure 4.37   Aluminium foil sheathed cable

The cables are stripped by scoring around the outer sheath with a knife. By bending the cable to and fro the aluminium breaks off. The tube that is left can now be pulled off leaving the inner cores.

The insulation of FP200 is made of silicon rubber, which is soft enough to be penetrated by the sharp edges of the aluminium sheath. To prevent this happening an insulated cap is fitted over the top of the remaining sheathing (Figure 4.38).

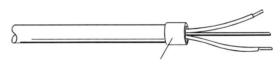

Insulated cap to protect cable insulation from the sharp edges on the aluminium sheath.

Figure 4.38

Where these cables have to be terminated into conduit accessory boxes a gland is used. This consists of a main body and locking bush which when tightened compresses a rubber "O" ring onto the cable (Figure 4.39).

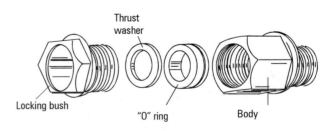

Figure 4.39

28
PVC/SWA cable termination

Objective: To terminate a PVC/SWA cable into a triple pole switch fuse to an acceptable standard.

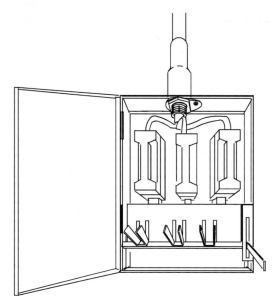

Suggested time: 1 hour

Materials:
1 measured length of PVC/SWA cable
1 complete gland and shroud
1 earth continuity connection
5mm brass nut and bolt
1 locknut
1 switch fuse

Points to be considered:

- was the gland assembled correctly?
- was the armour cut square?
- was the armour correctly placed in the gland?
- was the gland tightened?
- was the insulation on the conductors damaged?
- was enough cable left for the connections?
- were the conductors connected correctly?
- was the shroud fitted correctly?

29
Methods of conductor termination

Objective: To terminate control circuit conductors supplied by armoured control cable (with 5 or more cores) into metallic enclosures with lids, to an acceptable standard and test.

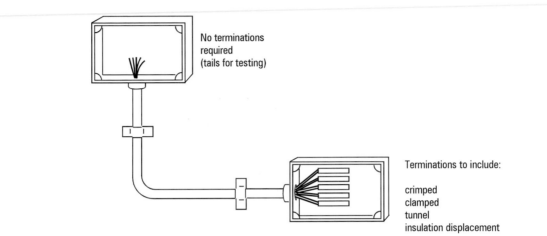

No terminations required (tails for testing)

Terminations to include:

crimped
clamped
tunnel
insulation displacement

Suggested time: 1 hour

Materials:
1 measured length of SWA cable in excess of 1 metre
2 proprietary glands
2 metallic enclosures 100 mm × 100 mm
crimped, clamped, tunnel and insulation displacement conductor terminations
cable supports

Points to consider:

- were the glands assembled correctly?
- was the armour cut square?
- was the armour correctly placed in the gland?
- was the gland tightened?
- was the insulation on the conductors damaged?
- was enough cable left for the connections?
- were the conductors connected correctly?

30
Aluminium foil sheathed cable

Objective: To terminate and install aluminium sheathed cables to form a lighting circuit and test.

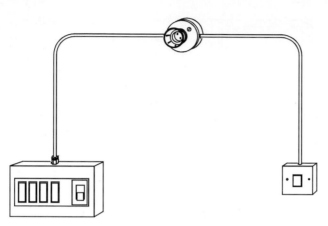

Suggested time: 2 hours

Materials:
1 metal consumer unit and protection
1 × one-way switch and pattress
1 lamp
green/yellow sleeving
1 measured length 1 mm^2 2 core and cpc PVC insulated aluminium sheathed cable
cable clips

Points to consider:

- was any of the cable insulation damaged?
- was the cable terminated into the consumer unit correctly?
- were all sheaths taken into the accessories?
- were all circuit protective conductors sleeved throughout their length?
- were all terminations tight?
- did the circuit test out correctly?
- did the circuit work?

5

Steel Conduit

On completion of this chapter you should be able to:

◆ cut and thread conduit ready for assembly
◆ bend conduit to a given dimension
◆ assemble a number of sections of conduit to form a complete assembly using boxes
◆ test the steel conduit for continuity
◆ draw in cables and connect up a lighting circuit

The tool kit required

For cutting, bending and threading

- hacksaw
- stocks and dies
- pipe vice
- bending equipment
- reamer
- files

For assembling

- pipe wrenches
- bush spanner
- screw driver
- hole punches and saws
- gland pliers
- spanners

For drawing in cables

- a draw tape
- side cutters
- pliers
- cable strippers

Power tools

- electric drill

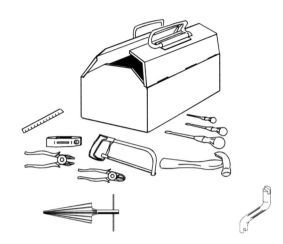

Figure 5.1

When steel conduit is installed correctly it provides a system of wiring that is both electrically and mechanically sound.

The steel tube provides a continuous circuit protective conductor that is strong and can withstand a great deal of mechanical abuse.

Figure 5.2   Standard thread on the end of the conduit

To maintain both the electrical and mechanical soundness, all connections and terminations must be made with tight clean joints. The most common method of jointing is by screwing the sections together (Figure 5.2). This requires the tube to be threaded on site and means all of the necessary equipment being available.

NOTE:
Even when a separate cpc is installed within a steel conduit one must ensure the conduit is electrically continuous throughout the system.

Stocks and dies

Threads are cut onto conduit using dies which are held in stocks (Figure 5.3). The thread is kept square to the tube by the guide on the front of the stocks.

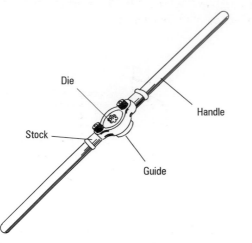

Figure 5.3   Stocks and dies

To cut a thread on a steel conduit we need to hold the tube firmly in a pipe vice. Figure 5.4 shows a pipe vice mounted on a tripod stand.

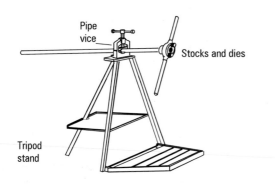

Figure 5.4

The threads need to be cut with care, as sound connections are required both mechanically and electrically. Lubrication such as cutting oil or tallow should be used whilst cutting threads and the dies should be kept clean and free from old cuttings and swarf. Threads are usually about the same length for couplings, and conduit fittings, but some special connections require at least twice the normal length of thread.

The length of the thread can be worked out using threaded couplings and connections. The conduits, in a straight through connection, should butt about half way (Figure 5.5).

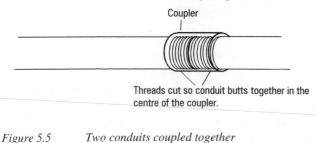

Figure 5.5   Two conduits coupled together

Sometimes it is not possible to turn the conduit to screw the sections together. In these situations a running coupler has to be used, as shown in Figures 5.6 and 5.7. A standard thread is used on the end of one of the conduits and on the other a thread long enough to take the complete length of a coupler plus a locknut. The two conduits are butted together and the coupler screwed back onto the standard length thread. The coupler is then "locked" into position using the locknut.

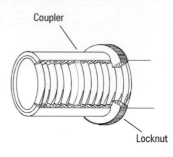

Figure 5.6   Running coupler ready for use

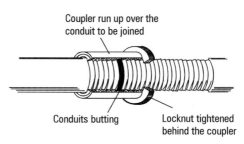

Figure 5.7   Running coupler installed

When terminating steel conduit into a spouted box the thread on the conduit should go right to the end of the thread provided in the box (Figure 5.8).

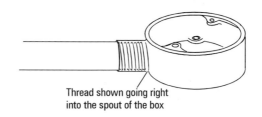

Figure 5.8

There should be no excess thread left over on the outside of the connection as this can corrode.

As the conduit system is only part of the installation and cables have to be pulled through the inside, all connections must be kept clean both inside and outside the tube. All rough edges and burrs must be reamed or filed away and any excess lubrication must be wiped off (Figures 5.9 and 5.10).

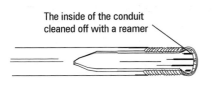

The inside of the conduit cleaned off with a reamer

Figure 5.9   *A section across the end of a piece of conduit*

Figure 5.10   *Conduit reamer*

Conduit supports

The conduits are held to the surfaces using a variety of saddles with either one or two fixing screws.

Spacer bar saddle

The most common saddle is the spacer bar as shown in Figure 5.11. The spacer bar is held to the surface with one fixing screw but the saddle is held with two set screws to the spacer bar. The holes in the saddle section are keyhole shaped to aid the person fixing it.

Figure 5.11   *Spacer bar saddle*

The conduit system

We must remember that the conduit is being installed to contain cables and so it must be constructed so that the cables can be pulled into the conduit without damaging them. The conduit is usually supplied in standard lengths of 3.75 m. In order to keep the stress on the cables when they are being installed to an acceptable level it is recommended that the maximum length between "cable draw in points" does not exceed 2 lengths of conduit.

Boxes used for switches or socket outlets are ideal as places to draw in cables (Figure 5.12). These can also be used to change the direction of the conduit run.

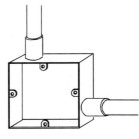

Figure 5.12

Where there are no specified outlets malleable conduit boxes are generally used (Figure 5.13). These are available in a wide range of spout configurations.

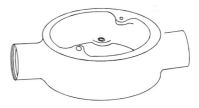

Figure 5.13

Conduit bends

There are several ways of going round corners with conduits. Special fittings can be used or the conduit can be bent.

Bending conduit

The minimum radius of a bend in conduit is determined from its overall diameter, as was the case with our cables (Figure 5.14).

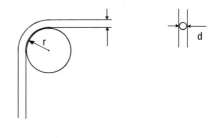

Figure 5.14   *Minimum radius of bend = 2.5 × diameter of conduit*

Machines can be used to make bends of the correct radius every time but other methods can be used which require more care. Even when using a bending machine (Figure 5.15), skill is required to get the bend in the correct place on the conduit.

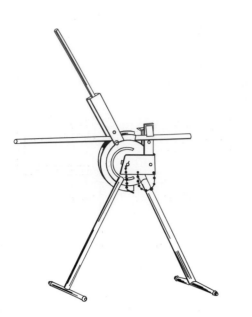

Figure 5.15   Bending machine

The wooden bending block

A wooden bending block is generally a $200 \times 100 \times 1000$ block of close grained wood which does not split easily. A hole, slightly larger than the conduit to be bent is drilled through about 100–150 mm from the end of the block at a slight angle to the timber (Figure 5.16). Each side of the hole is smoothed and shaped so that there are no sharp edges which would damage the conduit during the bending process.

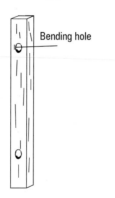

Bending hole

Figure 5.16   Wooden block

Making a good bend to set dimensions is a skill which takes practice. Bends made with a wooden bending block will usually be a greater radius than those made on a bending machine.

A long bend such as a right angle is made up of a number of small bends. The conduit is pushed through the hole in the bending block until the start of the intended bend. A small bend is made (Figure 5.17) and the conduit is pushed through the block about 20 mm and another small bend is made. This is continued until the required angle is achieved.

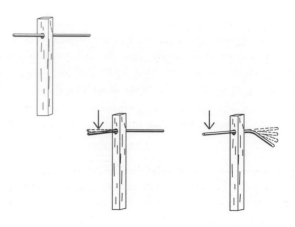

Figure 5.17

Circuit protective conductor

As the steel conduit may be used as the circuit protective conductor, when the conduit is assembled all joints must be made off tight. When the conduit is terminated into boxes the bushes should be tightened so that continuity is maintained. Bush spanners (Figure 5.18) are available to help with this.

Figure 5.18   Bush spanner

Where the metal box is painted this may have to be cleaned off as the paint may prevent a good electrical connection being made. Terminals are provided in the conduit boxes so that leads can be connected to provide a connection to accessories (Figure 5.19).

Remember

The same requirements exist even when a separate cpc is installed. This is because the conduit forms an "exposed conductive part" of the completed installation which must be earthed throughout its length.

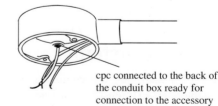

cpc connected to the back of the conduit box ready for connection to the accessory

Figure 5.19

Flexible steel conduit

Flexible steel conduit has a galvanised steel finish and is constructed in the form of a steel spiral. It can be cut to the required length by cutting through one section and then "unscrewing" the two pieces away from each other. In order to connect flexible steel conduit to the rigid conduit system, or to equipment, a gland or adaptor is screwed onto the spiral at each end.

Although it is made of steel this flexible steel conduit may not be used as a circuit protective conductor. A separate cpc must be installed inside the conduit with the other conductors to provide our earth continuity. This is true with all our flexible conduits regardless of their construction.

A typical application for flexible conduit is shown in Figure 5.20.

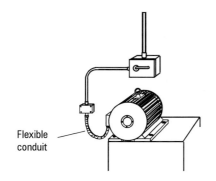

Figure 5.20

Wiring lighting circuits in conduit

Wiring in conduit is carried out using single core cables of the relevant colours. Unlike wiring with multi-core cables it is not necessary to use a 3 plate wiring system. A method very similar to the circuit diagrams where each line represents a single cable may be used (Figures 5.21 and 5.22).

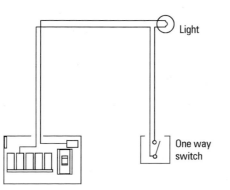

Figure 5.21   The cpc in this case is the steel conduit

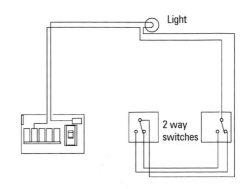

Figure 5.22   The cpc is the steel conduit

When cables are pulled into conduit care must be taken so that the insulation is not damaged.

If the cables have to be pulled more than a short distance a draw line should be used. This is usually a single cable pushed through first and the others pulled in on it. Over long distances a special tape may have to be used first to get round the conduit. The draw line would be pulled in on the tape and then the cables pulled in on the draw line.

By having the cables laid out as in Figure 5.23 they can be fed into the boxes without them crossing over or twisting up. When pulling cables in over a distance greater than an arm's length two people are required – one person to feed in the cables, the other to pull them through. Proprietary cable support equipment is available for use when pulling in cables. This is generally a purpose build arrangement, similar to that in Figure 5.23, that guides cables into the conduit system.

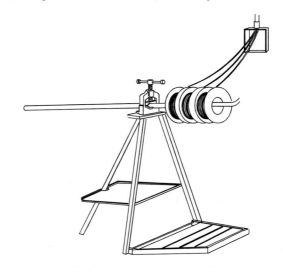

Figure 5.23   A tripod vice used to hold cable drums as cables are drawn into the conduit

Testing steel conduit as a circuit protective conductor

Where steel conduit or trunking is used as the circuit protective conductor it must be tested to ensure it is continuous. It must also have a low enough value in ohms to allow large currents to flow in the event of a fault so that the protective device can operate within the required time.

31
Connecting a 3-phase supply socket outlet

Objective: To connect to a 3 phase supply to a socket outlet via flexible conduit and armoured cable.

Complete the installation from a TP & N isolator to a TP & N direct-on-line starter using armoured cable and connect to a 3 phase socket outlet via a flexible conduit as shown. Test the installation.

Suggested time: 2 hours

Materials:
- 1 TP & N isolator
- 1 measured length of 4 core SWA cable (min. 2 metres)
- Proprietary glands and shrouds for SWA
- 1 direct-on-line starter
- 1 measured length of flexible conduit
- Proprietary glands for flexible conduit
- Crimped spade terminations
- measured lengths of 2.5 mm^2 single PVC red, yellow and blue cables
- 1 measured length of 2.5 mm^2 single PVC green/yellow cable
- earth continuity ring
- brass nuts and bolts

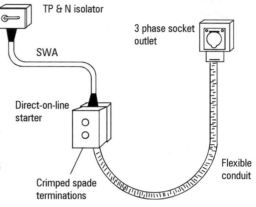

Points to consider:
- were the terminal connections made off correctly?
- were the crimped spade terminations connected correctly?
- was the cable connected correctly into the 3-phase socket outlet?
- did the circuit test out correctly?

Armoured cable requirements:
- was the termination correct?
- was the earth connected to the main earth terminal?
- was the armour contained within the gland?

32
Threading and bending steel conduit

Objective: To bend and thread a short length of conduit as shown.

Suggested time: ¾ hour

Materials:
- 1 measured length of 20 mm steel conduit

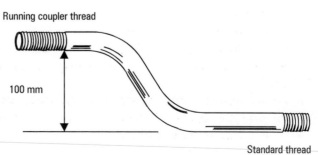

Points to consider:
- were the inside burrs removed from the conduit?
- were all sharp edges removed from the outside of the conduit?
- were the threads cut to an acceptable length?
- were the overall dimensions achieved?

33
Steel conduit installation

Objective: To cut, thread and bend steel conduit as shown and test for continuity.

Suggested time: 2 hours

Materials:
- 1 measured length of 20 mm steel conduit
- 1 through box
- 2 switch boxes
- 1 consumer unit box
- 1 proprietary manufactured 90° bend or elbow
- conduit supports

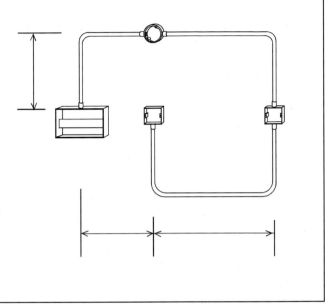

Points to consider:
- were the inside burrs removed from the conduit?
- were all sharp edges removed from the outside of the conduit?
- were the threads cut to an acceptable length?
- were all connections tightened?
- were the overall dimensions achieved?

34
Wiring a steel conduit installation

Objective: To wire a two way lighting circuit in the steel conduit installation from the previous exercise.

Suggested time: 1½ hours

Materials:
- 1 measured length of 1.00 mm² single PVC sheathed black cable
- 1 measured length of 1.00 mm² single PVC sheathed red cable
- 1 × 2 plate batten holder
- 2 two-way plate switches
- 1 consumer unit interior

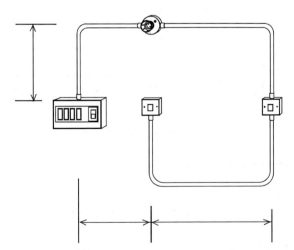

Points to consider:
- were all the terminations made off correctly?
- does the insulation of the cables go up to the terminations?
- were the correct colour cables used throughout the installation?
- are the correct fuseway and neutral bar termination used in the consumer unit?
- has a suitable amount of spare cable been left at the accessories and consumer unit?
- is the circuit connected correctly?
- does the circuit work?

35
Complete steel conduit installation

Objective: To tube and wire a complete lighting circuit consisting of 2 one-way switches controlling two lights and test.

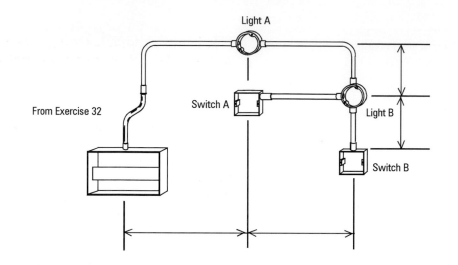

Note:

Switches A and B are one-way switches.
Switch A controls light A and switch B controls light B

Suggested time: 4 hours

Materials:
- 1 measured length of 20 mm steel conduit
- 1 through box
- 1 "T" box
- 2 switch boxes
- 1 complete consumer unit
- 2 × 2 plate batten holders
- 2 one-way plate switches
- 1 measured length of 1.00 mm^2 single PVC sheathed black cable
- 1 measured length of 1.00 mm^2 single PVC sheathed red cable
- couplers and bushes

Points to consider:
- was the conduit installed level and fixed to an acceptable standard?
- were the overall dimensions achieved?
- was the circuit wired to an acceptable standard?
- were the conduit ends cut square and have they had the burrs removed?
- were the conduits threaded correctly?
- were the correct fuseway and neutral bar termination used in the consumer unit?
- was a suitable amount of spare cable left at accessories and consumer unit?
- was the circuit wired correctly?
- does the circuit work?

6

Rigid PVC Conduit

Some of the installation methods used in steel conduit apply to PVC conduit. However, there are also several differences.

Similarities

- Although made of PVC the conduit boxes and accessory outlets are similar.
- Bending radii are the same.
- Wiring methods are similar.
- The same diameter tubes are used.

Differences

- Circuit protective conductors have to be installed as the conduit is not a conductor.
- Bends are made using a spring similar to that used by plumbers for copper pipe.
- Fixings need to be closer as PVC is not as strong as steel.
- Allowances must be made for expansion as this is 6 times as great in PVC as in steel.
- Terminations and joints are not threaded and screwed but glued.
- Bends have to be overbent as they have a tendency to open out after a short time.

On completion of this chapter you should be able to:

- ◆ cut and bend rigid PVC conduit to given dimensions
- ◆ make a right angle bend to an acceptable radius
- ◆ cut the tube square
- ◆ de-burr the tube whenever necessary
- ◆ complete a small installation to an acceptable standard
- ◆ wire a ring final circuit in conduit using single cables

The tool kit required

For cutting and bending
- junior hacksaw
- bending spring
- cutting tool

For assembling
- PVC solvent
- screw drivers
- bush spanner

For drawing in cables
- a draw tape
- side cutters
- pliers
- cable strippers

For testing
- instruments

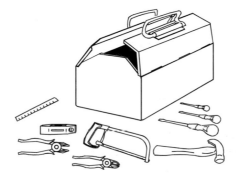

Figure 6.1

Bending PVC conduit

It is very easy to flatten PVC conduit when trying to bend it. A spring is used inside the conduit to help to stop this happening (Figure 6.2), but even so care still has to be taken. The usual method of actually bending the conduit is round the knee (Figure 6.3). As most knees are well under the accepted radius a number of short bends have to be made to create a longer one. Bends should always be made overbent; they tend to straighten out.

Figure 6.2

Figure 6.3

Terminating PVC conduit

Where PVC conduit goes into a spouted conduit box the conduit is glued into the spout (Figure 6.4).

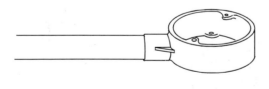

Figure 6.4

The glue melts the two surfaces to be connected and makes a welded connection.

The selection of conduit boxes for PVC installation is similar to that for steel conduit and is shown in Figure 6.5.

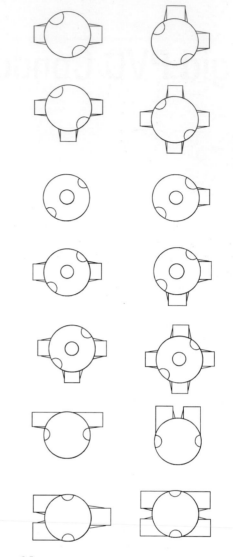

Figure 6.5

In addition to the circular boxes there are also inspection fittings (Figures 6.6–6.8). These do not allow the same room for pulling in cables but can be used in confined spaces.

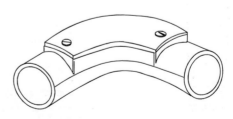

Figure 6.6   Inspection elbow

Figure 6.7   *Inspection bend*

Figure 6.8   *Inspection tee*

Where lengths of conduit have to be connected together a coupler is used. There are two types of these. One which is glued to each conduit to be joined (Figure 6.9) and one which is glued one end, the other being sealed with a semi-permanent mastic to take up any expansion (Figure 6.10).

Figure 6.9   *Coupler glued to each conduit*

Figure 6.10   *Coupler glued one end and sealed with a semi-permanent mastic on the other*

Terminations into accessory boxes (Figure 6.11) can be made in a variety of ways, one of which is similar to that for steel conduit. A coupler and male bush can be used where the coupler is glued to the end of the conduit and then a male bush is tightened into the threaded section.

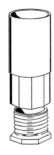
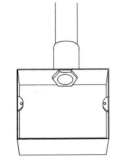

Figure 6.11   *Termination into an accessory box*

Figure 6.12 shows a male adaptor which is used with a locking nut.

Figure 6.12   *Male adaptor*

Various methods of clip in connection can be used, one of which is shown in Figure 6.13.

Figure 6.13   *Clip in connection*

Fixing PVC conduit

Fixing devices for PVC conduit are similar in principle to those used for steel conduit (Figure 6.14).

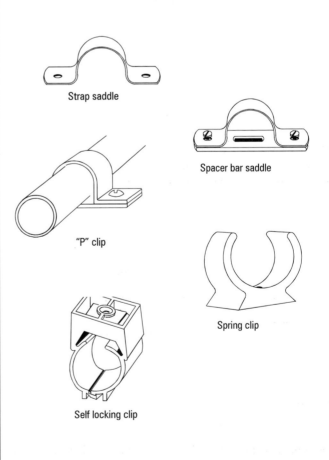

Strap saddle

Spacer bar saddle

"P" clip

Spring clip

Self locking clip

Figure 6.14   *Fixing devices for PVC conduit*

Wiring ring final circuits in conduit

When wiring any ring final circuit it is essential to ensure that a ring is maintained in all conductors. With twin and cpc cable the three conductors are always automatically taken together. When single cables are used in conduit each conductor must run the same route as the other conductors and form a ring. In PVC conduit this also applies to the circuit protective conductor. Where a number of socket outlets are connected to the same conduit as shown in Figures 6.15 and 6.16 both parts of the ring are run through the same conduit.

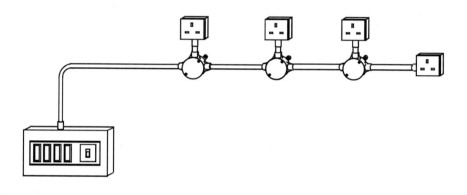

Figure 6.15   Four socket outlets wire as a ring final circuit in rigid PVC conduit

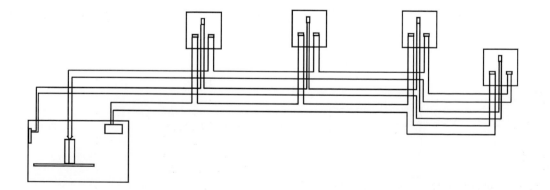

Figure 6.16   Circuit diagram of the ring final circuit

36
PVC conduit bend and double set

Objective: To cut and bend a measured length of PVC conduit as shown.

Suggested time: 30 minutes

Materials:

1 measured length of rigid PVC conduit

Points to consider:

- were the inside burrs removed from the conduit?
- were the bends to an acceptable radius?
- was the conduit bent without twisting?
- was the conduit bent without ripples or kinks?

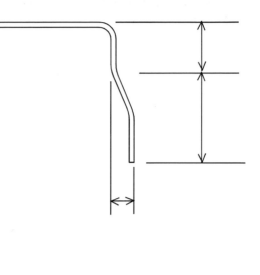

37
Rigid PVC conduit for radial circuit

Objective: To cut, bend and assemble rigid PVC conduit as shown in the diagram.

Suggested time: 45 minutes

Conduit marked "x" is that produced in an earlier task.

Materials:

1 consumer unit box
1 socket box
1 through box
1 measured length of rigid PVC conduit

Points to consider:

- was the right angle bend formed to an acceptable radius?
- were the conduit ends cut square and were the burrs removed?
- were the terminations made to an acceptable standard?
- were the overall dimensions correct?

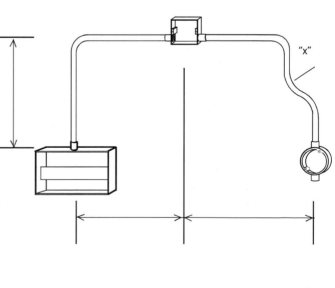

38
Radial circuit in rigid PVC conduit

Objective: To use the rigid PVC conduit installation produced in the previous task to wire a radial circuit. Test for continuity and insulation resistance.

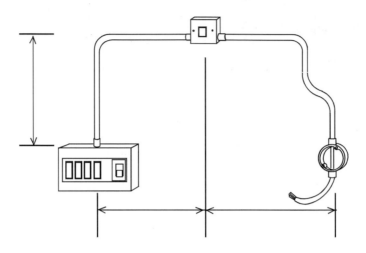

Suggested time: 1 hour

Materials:

consumer unit interior with protective device
1 × 20 A double pole switch
1 measured length of 1.0 mm² black cable
1 measured length of 1.0 mm² red cable
1 measured length of 1.0 mm² yellow/green cable
3 × 5 A connector blocks
insulation resistance and continuity tester

Points to consider:

- are all cables connected correctly?
- is the correct polarity observed?
- are all terminations made tight?
- were the correct fuseway and neutral bar termination used in the consumer unit?
- was a suitable amount of spare cable left at accessories and consumer unit?
- was the correct procedure used to verify continuity of the cpc?
- were appropriate precautions and checks carried out prior to the insulation resistance test?
- were acceptable measured values obtained?

39
Rigid PVC conduit socket installation

Objective: To cut, bend and assemble rigid PVC conduit as shown.

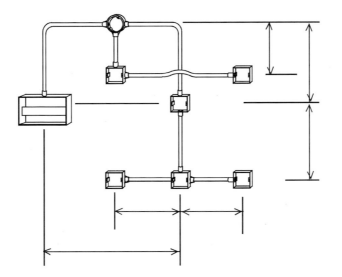

Suggested time: 2 hours

Materials:

1 measured length of 20 mm rigid PVC conduit
1 "T" circular box
6 socket outlet boxes
1 consumer unit box
1 proprietary manufactured 90° bend
2 different adaptor types

Points to consider:

- were the inside burrs removed from the conduit?
- were all of the conduits terminated correctly?
- were the bends made to an acceptable radius?
- were the overall dimensions achieved?
- was the conduit set symmetrically over the vertical conduit obstruction?
- was the conduit bent without ripples or kinks?

40
Wiring a rigid PVC conduit ring circuit installation

Objective: To wire the conduit installed in Exercise 39 with 3 socket outlets connected as a ring final circuit, one fused spur supplying a flex outlet and one non-fused spur, as shown. Test for continuity of ring final conductors, insulation resistance and polarity. Energise the circuit and test for correct operation.

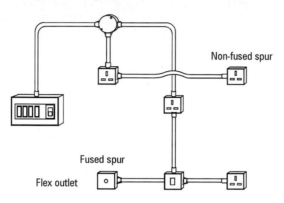

Non-fused spur

Fused spur

Flex outlet

Suggested time: 2 hours

Materials:

4 x 13A socket outlets
1 measured length 2.5 mm^2 single PVC red cable
1 measured length 2.5 mm^2 single PVC black cable
1 measured length 2.5 mm^2 single PVC yellow/green cable
1 consumer unit interior
1 conduit PVC box lid
1 fused spur
1 flex outlet
insulation resistance and continuity tester

Points to consider:

- are all terminations made off correctly?
- does the insulation of the cables go up to the terminations?
- are the cables connected to the correct polarity?
- was the correct procedure used to verify continuity of the cpc?
- were appropriate precautions and checks carried out prior to the insulation resistance test?
- were acceptable measured values obtained?
- is the circuit connected correctly?
- does the circuit work?

7

Metal Trunking

On completion of this chapter you should be able to:

◆ cut the trunking square
◆ cut and fabricate metal trunking to given dimensions
◆ fabricate a right angle bend to given dimensions
◆ fabricate a "T" junction to given dimensions
◆ couple conduit to trunking using recognised methods

The tool kit required
- hacksaw
- hammer
- screw driver
- spanners
- pop rivet tool
- hole saw
- cone cutter
- hole punch
- centre punch
- flat file
- round file

Power tools
- electric drill

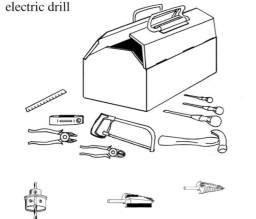

Figure 7.1

Where a number of cables have to be installed through the same route, trunking (Figure 7.2) offers a good alternative to conduit. Trunking provides access to the cables throughout the whole length making it easier to install and alter cables.

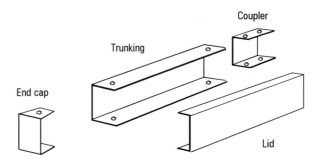

Figure 7.2

Trunking systems are fabricated on site and although standard fittings are available it is sometimes necessary to make shapes to fit the structure of the building.

When making bends and junctions the bending radius of the cables must be considered. Large cables require a suitable bending radius and the bends in the trunking should reflect this. When smaller cables are used a tighter trunking bend can be fabricated. As with steel conduit the metal enclosure may be used as the circuit protective conductor. In any event the electrical and mechanical continuity must be maintained. With some manufacturers of trunking this may require copper connecting strips across the joint to ensure continuity.

When cutting trunking a burr is formed. This must be filed off to prevent the cables being damaged.

Cutting trunking square

To cut trunking square it must first be marked off using a square and scriber (Figure 7.3). Before cutting, wooden blocks should be fitted inside the trunking to help it keep its shape. A hacksaw is generally used to cut the trunking and both hands should be on the saw keeping it firm. If your head is kept behind the saw with your eye looking down onto the line, the cut should be kept square.

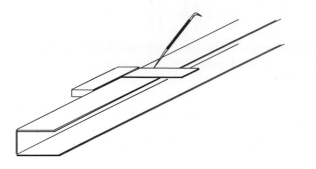

Figure 7.3

Fabricating a 90° angle

Mark out 45° on each side of the centre and cut out this section (Figure 7.4).

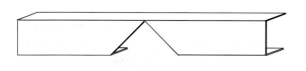

Figure 7.4

Bend down the side (Figure 7.5).

Figure 7.5

To ensure a good right angle without distorting the trunking, a block of wood should be placed inside the trunking.

Rivet or bolt a strengthening piece across the joint at the back of the trunking (Figure 7.6).

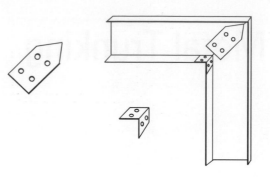

Figure 7.6

Rivet or bolt a strengthening piece over the edge so that it will also protect the cable from sharp edges. The lid can be cut in two sections mitred at 45° so that they meet across the corner.

Fabricating a double set

A double set consists of two bends in opposite directions. Each bend has one side that is not cut but bent. To help with the marking out it is often useful to chalk out the finished shape on a flat surface and lay the straight trunking on to it.

Mark and cut out the two sections (Figure 7.7). So that the trunking does not distort, wooden blocks should be placed inside.

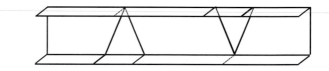

Figure 7.7

As with the 90° bend, strengthening pieces are riveted or bolted at the back of the trunking (Figure 7.8).

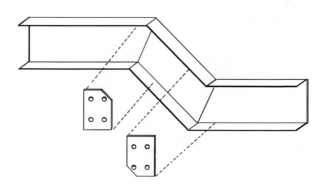

Figure 7.8

The cut edge should have a strengthening piece which also covers the sharp edges and protects the cables (Figure 7.9).

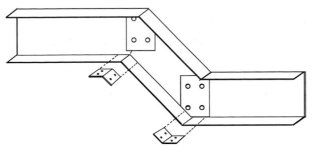

Figure 7.9

The lid is cut into three sections so that they butt at the correct angles across the corners (Figure 7.10).

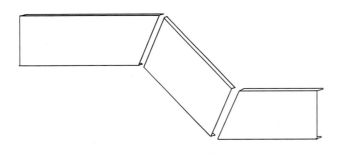

Figure 7.10

Fabricate a "T" junction

A "T" junction is made out of two separate pieces of trunking joined together. The top section of the "T" has to have a section removed from the centre at one side (Figure 7.11).

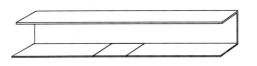

Figure 7.11

The other section has the sides cut down and bent over so that both sides are the same (Figure 7.12).

Figure 7.12

The two sections are fitted together and riveted or bolted. All of the edges must be filed so that cables cannot be damaged when they are installed (Figure 7.13).

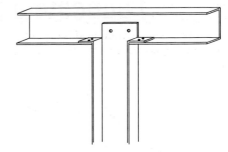

Figure 7.13

The lid is made in two sections, one across the top, the other butting up to it. The section across the top must have the lip removed where it goes across the vertical section.

Terminations into trunking

Conduits are often used to or from steel trunking and a standard termination of male bush and coupler is used (Figure 7.14).

Figure 7.14

Where paint on the trunking is a problem for continuity a shake-proof washer may be used under the male bush to penetrate into the steel through the paint.

The holes in the trunking for conduit entries may be produced a number of different ways using hole saws, hole punches and cone cutters.

Cone cutter

This consists of a hardened steel cone shaped drill (Figure 7.15). The size of the hole is dependent on the depth the cutter is allowed to go. The stepped cutter helps with this problem.

Hardened steel cone shaped drill

Stepped cutter

Figure 7.15

Care must be taken not to go too far and make a hole too large.

41
Fabricate a 90° bend in trunking

Objective: To fabricate a 90° bend and lid from a straight length of steel trunking and complete the assembly as shown using a proprietary manufactured bend and coupler.

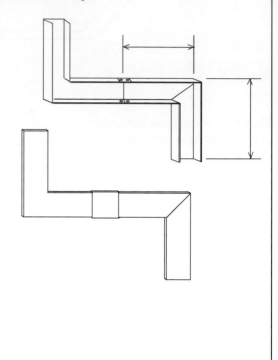

Suggested time: 2 hours

Materials:

1 measured length of trunking
1 proprietary manufactured bend
1 coupler
1 copper continuity link

Points to consider:

- was the trunking bend made off square?
- were the cut edges filed smooth?
- were the strengthening pieces fitted correctly?
- was the lid cut to fit the trunking to an acceptable standard?
- was the completed assembly to an acceptable standard?
- was the continuity link installed correctly?

42
Double set in steel trunking

Objective: To fabricate a double set in steel trunking to the details as shown.

Suggested time: 2 hours

Materials:

1 measured length of trunking and lid
bolts or rivets

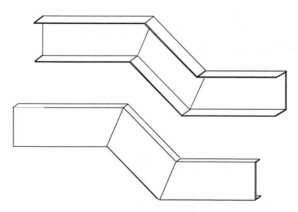

Points to consider:

- were the bends made so the two end sections are parallel?
- were the cut edges filed smooth?
- were the strengthening fitted correctly?
- was the lid cut to fit the trunking to an acceptable standard?

43
"T" junction in steel trunking

Objective: To cut one piece of trunking and lid into two and form them into a "T" junction as shown.

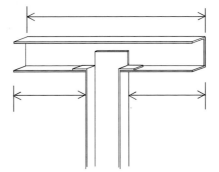

Suggested time: $1\frac{1}{2}$ **hours**

Materials:

1 measured length of trunking and lid
bolts and rivets
1 copper continuity link

Points to consider:

- were the ends cut square?
- was the centre section cut out cleanly?
- was the joint made square?
- were all edges filed clean?
- was the lid cut to fit the trunking to an acceptable standard?
- was the "T" junction made to the correct dimensions?
- was the continuity link installed correctly?

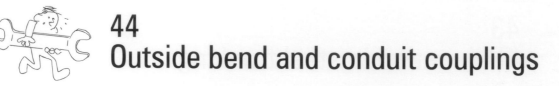

44
Outside bend and conduit couplings

Objective: To fabricate an outside bend in trunking and couple a conduit box to it as shown.

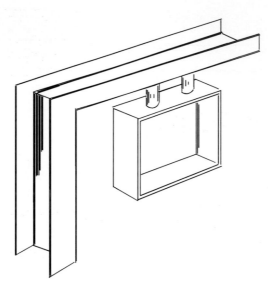

Suggested time: $2\frac{1}{2}$ **hours**

Materials:

1 measured length of trunking
1 adaptable conduit box
2 conduit couplers
4 conduit male bushes

Points to consider:

- were the ends cut square?
- was the joint made square?
- were the edges filed?
- was the bend made to the correct dimension?
- were the holes for the couplers cut clean?
- were the holes made in the correct places?
- were the couplings to the conduit adaptable box made to an acceptable standard?

8

Cable Tray

On completion of this chapter you should be able to:

◆ cut cable tray square
◆ fabricate a 90° bend to an acceptable standard
◆ fabricate a "T" junction to an acceptable standard
◆ fabricate a reduction section to an acceptable standard
◆ make inside and outside bends using acceptable methods
◆ fabricate a bracket to an acceptable standard

The tool kit required

• hacksaw
• hammer
• files
• screw driver
• spanners

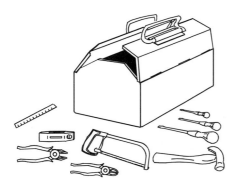

Figure 8.1

Cable tray (Figures 8.2 and 8.3) is used as a method of support for cables. It consists of lengths of perforated tray which has lips or flanges on the edges. This is generally manufactured in galvanised steel, but is available with specialist coatings in plastic. For the practical exercises we shall be considering the standard glavanised tray.

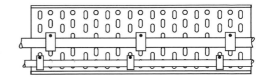

Figure 8.2

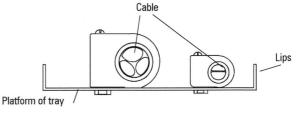

Figure 8.3   *Section through cable tray*

Manufactured bends and junctions are available (Figure 8.4).

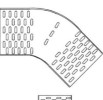
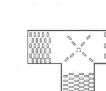

Figure 8.4

Fabricating cable tray

It is often necessary to fabricate shapes on site and when making bends the bending radius of the cables that are to be installed on the tray should always be considered.

Fabricating a flat 90° bend producing a sharp angle

This is suitable for cables with a relatively small diameter so that the radius of the bend does not have to be too great.

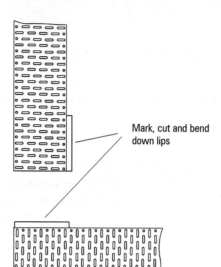

Mark, cut and bend down lips

Figure 8.5

The bend is made from two sections of cable tray laid over each other (Figures 8.5 and 8.6). The side lips have to be cut and folded flat so that they do not present an obstruction.

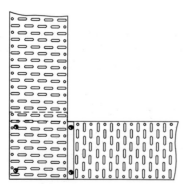

Figure 8.6

Fabricating a flat 90° bend producing a wide angle

This bend is used where the bends in the cables require a larger radius.

This can be fabricated out of one piece of cable tray using the method shown in Figures 8.7 and 8.8.

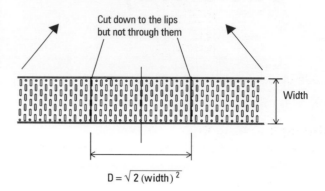

Cut down to the lips but not through them

Width

$$D = \sqrt{2 \, (\text{width})^2}$$

Figure 8.7

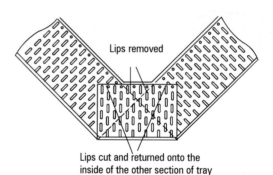

Lips removed

Lips cut and returned onto the inside of the other section of tray

Figure 8.8

Forming a "T" junction

A "T" junction is made from two separate pieces of cable tray (Figure 8.9).

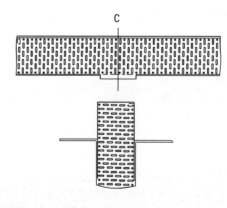

C

Figure 8.9

On the top piece of tray a section of the lip is removed the width of the tray to be joined. The lower piece of tray is prepared by cutting the lip away from the tray and bending it back.

When the "T" is complete the side lips fit to those of the top section with rivets or bolts (Figure 8.10).

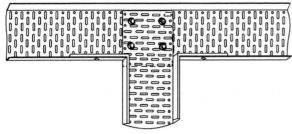

Figure 8.10   Completed "T" junction

Forming a reduction

It is important to maintain the flanges of the tray throughout the reduction.

The reduction is made by marking out the smaller tray onto the larger (Figure 8.11).

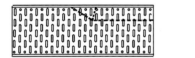

Figure 8.11

This can be cut but it should be left connected by the lip (Figure 8.12).

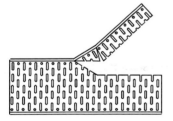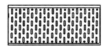

Figure 8.12

The cut section can now be overlapped ready for bolting together (Figure 8.13).

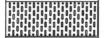

Figure 8.13

The smaller tray can now be overlapped over the reduction and bolted to it (Figure 8.14).

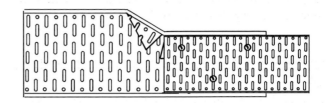

Figure 8.14

Inside and outside bends

Both outside and inside bends can be made using a bending machine. However there are also ways of fabricating them by hand.

Fabricating an outside bend

The lip of the tray limits the tray from bending outwards. As the lip will not stretch under normal pressure it has to be cut. The sharper the bend has to be the more cuts are required (Figure 8.15).

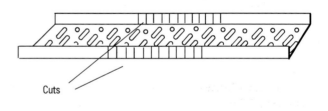

Cuts

Figure 8.15

To produce an even bend the cuts should be evenly spaced. After the bend has been made (Figure 8.16), the cut edges should be filed to reduce the risk of damage to the cables.

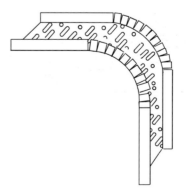

Figure 8.16

Fabricating an inside bend

An inside bend has the opposite problem to that of an outside bend. With the inside bend the lip needs to be shorter. A simple tool, as shown in Figure 8.17, can be made to help with this problem.

Figure 8.17

By making a number of small crimp bends in the lip the tray will take up a radius. The crimps should be evenly spaced along the sides and each side must be kept even (Figure 8.18).

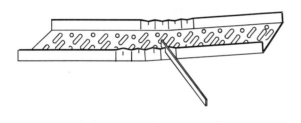

Figure 8.18

After crimping a few bends on one lip an equal number at the same spacing should be carried out on the other (Figure 8.19). The closer together the crimps the tighter the bend, the further apart the wider the bend.

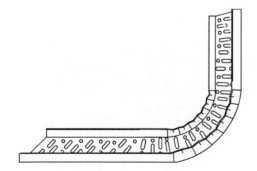

Figure 8.19

In practice both these bends can also be produced using a proprietary cable tray bending machine.

Making a cable tray bracket

There are a number of different types of bracket for mounting cable tray. The one shown in Figure 8.20 can be varied very simply to apply to many situations.

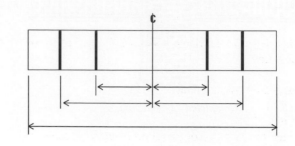

Figure 8.20

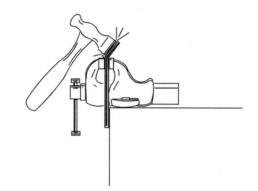

Figure 8.21

A section of mild steel bar should be marked off where the bends are to be, as shown in Figure 8.20. The bar can be bent in a vice using a hammer (Figure 8.21). Care must be taken as the metal may be springy and the hammer may bounce back.

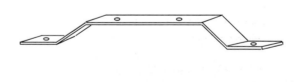

Figure 8.22

The bends need to be even so that the platform of the bracket is parallel to the fixing surface and the fixing feet lay flat (Figure 8.22).

45
Fabricate a 90° bend in cable tray

Objective: To fabricate a sharp 90° bend using two pieces of cable tray.

Suggested time: 1 hour

Materials:
1 measured length (or two short lengths) of cable tray
nuts, bolts and rivets

Points to be considered:

- was the bend made to 90°?
- were the edges sharp or have they been filed off?
- was the bend made strong enough?

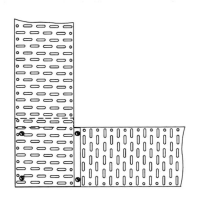

46
"T" junction in cable tray

Objective: To fabricate a "T" junction out of cable tray as shown in the diagram.

Suggested time: 1½ hours

Materials:
1 measured length (or two short lengths) of cable tray
nuts, bolts and rivets

Points to be considered:

- was the "T" made to 90°?
- were the edges sharp or have they been filed off?
- was the "T" joint made strong enough?

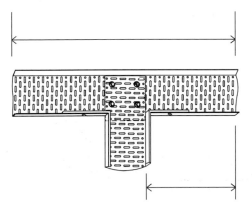

47
A reduction section in cable tray

Objective: To fabricate a reduction section of cable tray as shown in the diagram.

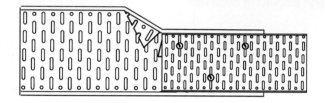

Suggested time: $1\frac{1}{2}$ hours

Materials:
1 measured length of large cable tray
1 measured length of smaller cable tray
nuts and bolts

Points to be considered:

- was the reduction made to the correct dimensions?
- are the edges sharp or have they been filed off?
- was the reduction made strong enough?

48
A double set in cable tray

Objective: To produce double sets in a measured length of cable tray, by making inside and outside bends.

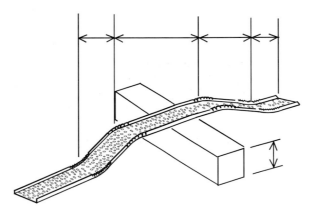

Suggested time: 2 hours

Materials:
1 measured length of cable tray

Points to consider:

- were the inside and outside bends made to an acceptable standard?
- were the edges sharp or have they been filed off?
- was the double set made to the correct dimensions?
- have any unwanted twists been produced during the fabrication?

49
Cable tray combined system

Objective: To assemble the results of the previous cable tray exercises to produce a complete system.

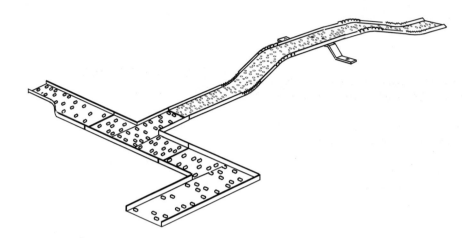

Suggested time: $1\frac{1}{4}$ hours

Materials:
Completed previous cable tray exercises
short lengths of cable tray for joints
1 measured length of mild steel bar

Points to be considered?

- were the edges left sharp or have they been filed off?
- were all bolts tightened?
- do all sections match where they should?

9

Associated Electronics

On completion of this chapter you should be able to:

◆ carry out soldering exercises to acceptable standards
◆ disconnect and replace components on a printed circuit board
◆ carry out tests on resistors, capacitors and semiconductors using appropriate test instruments
◆ carry out electrical connections by terminating coaxial cables
◆ terminate a ribbon cable into an insulation displacement connector to an acceptable standard

Shown in the photograph below are some of the general tools you will come across when working with electronic components. Remember to keep all tools in a good condition ready for use. Always choose the correct tool for the job. For instance the use of installation pliers to bend and insert a 0.4 W resistor into a PCB is obviously wrong!

An electrician's tool kit is likely to contain basic items such as lightweight pliers, cutters and screwdrivers suitable for use with electronic equipment. Some examples are shown in Figure 9.1.

The tool kit required
For assembling
• lightweight pliers
• miniature side cutters
• wirestrippers
• miniature screwdrivers
• soldering iron
For testing
• ohmmeter
• multimeter

Electricians' pliers compared in size to those required for use with electronic components.

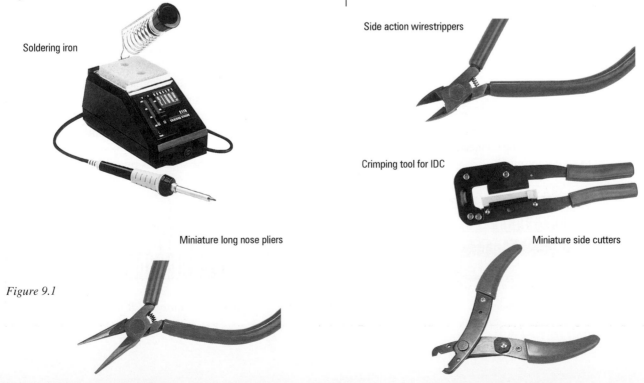

Soldering iron

Side action wirestrippers

Crimping tool for IDC

Miniature long nose pliers

Miniature side cutters

Figure 9.1

Cables

The types of cables generally available for electronics work can be grouped as:

- equipment – for example single core wire in 7/0.2 mm size with a variety of coloured insulation usually PVC
- data transmission – for example overall screened twisted pair suitable for the interconnection of computer equipment
- audio visual – SCART audio visual cable
- signal and telephone – for example security/alarm cable or IDC flat ribbon cable
- coaxial – for example television aerial download

Connections

Connections fall into two categories
- readily separable by unplugging (connectors)
- mechanically fixed such as soldered and wire wrapped

Connectors

Plugs and sockets

There are many types of plugs and sockets that are used in electronics to connect sections and equipment together. These range from a single pin and socket arrangement to complex multipin and screened connections.

A jack socket and plug (Figure 9.2) has many applications in audio equipment. The cable connections may be terminal screw or solder.

Figure 9.2   Jack socket and plug

So as to ensure good electrical connections the contacts are often gold-plated silver. The cable connections to this type of plug are made on to solder terminals.

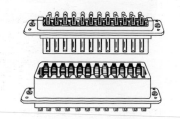

Figure 9.3   24-way module connector

Coaxial sockets and plugs (Figure 9.4) and other similar connections are used on screened cables where the screening is connected to the outside metal and the core to the centre pin.

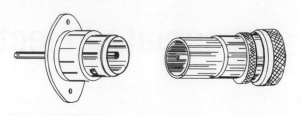

Figure 9.4   Coaxial socket and plug

When removing multiple connectors that may be interchangeable always note where they were removed from so that they can be replaced correctly. Although the connections are usually soldered some of the latest sockets and plugs have a screw-in connection designed for use where facilities for soldering are not easily available.

Connecting to printed circuit boards

A printed circuit board (PCB; Figure 9.5) is an insulated base with tracks of conducting material running in an intricate pattern over one or both sides of the board. There are also multi-layer boards possibly up to as many as 10 layers.

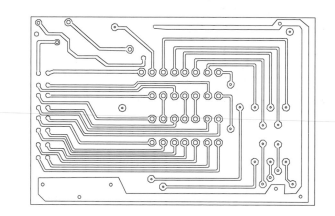

Figure 9.5   Printed circuit board

Where printed circuit boards are made as a replaceable unit they often incorporate a plug-in arrangement on the board. One example of this is the edge connector (Figure 9.6). Here the printed circuit board has been made so that the track goes out to the edge of the board. To ensure a good electrical contact the area that plugs into the socket is often gold-plated.

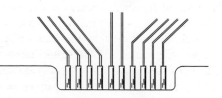

Figure 9.6   Edge connector on PCB

Printed circuit boards should always be handled with care. The manufacturing process of the copper tracks means that they may be very thin in places and can easily be damaged. Consideration must also be given to the fact that many of the components on the boards may be damaged by the minute voltages in the human body. The boards must be handled by their edges without touching the conductor tracks. It may be necessary to wear an earthed strap to stop the effects of static electricity from damaging the sensitive components.

Circuit components or their mounting bases are connected directly onto the board by soldering. Great care must also be exercised when replacing the components which have been mounted on a PCB. The copper tracks are very fine and can be fractured if the board is bent, flexed or overheated during the process.

A matrix/stripboard is a synthetic resin bonded paper (SRBP) panel that has 1 mm diameter holes arranged in a 0.1 inch grid. Boards, which may have copper strips bonded to one side, are available in various types and sizes and are particularly useful for prototype work. They accept terminal pins, as shown in Figure 9.7, which provide a post to which connecting wire and components can be soldered.

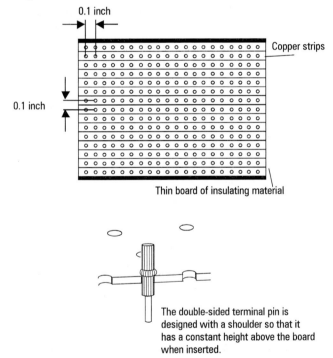

Figure 9.7   Copper strips on an insulating board with a matrix of 0.1 inch holes and a terminal pin.

The components are inserted into the non-copper side of the board and the leads of the components are cut short before soldering to the copper strips. These copper strips take the place of the wire links that would be used on a plain matrix board. The copper strips are continuous but can be broken using a small drill or strip cutter. The drill is placed on the hole where the break is to be made and then turned a few times manually until the very thin copper strip is removed.

Copper board is more expensive than plain board because of the additional cost of the copper. If plain board is used then links must be made between the components on the reverse side of the board.

Mechanically fixed components

Connector blocks
Connector blocks provide a means of connection and disconnection of cables from a PCB. The blocks can be single or multi-way versions and the wires can be secured by screws, clamps or soldering depending on the type and the reason for the connection. A range of connector block sizes is available to suit the gauge of wire. Wires attached by screws facilitate easy disconnection when servicing, testing or replacing whereas soldered connections are suitable for more permanent installations.

Most types of screw connector block feature wire guards under each screw to protect the wire from any twisting or cutting action by the screw.

Insulation displacement connection (IDC)
There are methods of connection which do not require the removal of the insulation from a cable.

The connection is made by pressing the cable onto a "V" shaped blade (Figure 9.8). This blade pushes through the insulation and then touches onto the conductor. The blade is shaped so as not to damage the conductor.

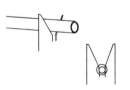

Figure 9.8

These are used extensively for ribbon cables (Figure 9.9) where all of the conductors can be connected at the same time with the minimum separation of individual cores.

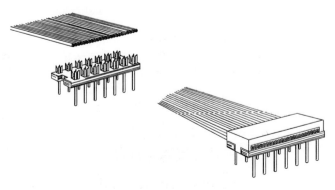

Figure 9.9   Dual-in-line (DIL) IDC header

To ensure good electrical and mechanical connections special tools have been developed for inserting the cables into the connectors. Figure 9.10 shows a typical handheld single core tool.

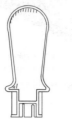

Figure 9.10

Where there are a lot of connections to be made, more sophisticated multicore press tools are available for commercial volume production. For prototype or repair work hand tools are available for crimping multi-way connectors. One, a universal type vice jaw tool (Figure 9.11), holds the connector and the cable in position in a bench vice. When the jaws of the vice are closed the outer insulation is pierced (you will normally hear a "click") and the connection is made to the inner conductors. A strain relief clip is usually fitted to complete the termination. Any excess ribbon cable should be trimmed off neatly using cutters or a sharp knife.

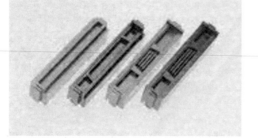

Figure 9.11   Universal vice jaw tool

Crimp connections

Crimped connections are carried out using a purpose made crimping tool (Figure 9.12). It connects an individual wire to a contact which is inserted into a connector housing. Often there is a separate tool available for removing the contact from the connector body and it is likely that you will find that manufacturers' will have their own specific tool for their own products.

Figure 9.12   Crimping tool for IDC connectors

There are other hand tools available (Figure 9.13), some of which are specific to manufacturers, and these may only be used with the appropriate connectors.

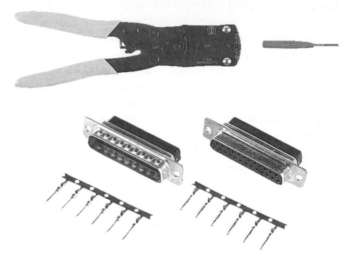

Figure 9.13   Crimp tool and connectors

Solder connections

The most used connection is the soldered one. This generally uses an alloy of tin and lead to bond metallic materials together. The alloy used for electrical soldering is 60% tin, 40% lead and this melts at 188 °C. To ensure a good electrical and mechanical connection a resin flux is used to prevent oxidation of the metals. The solder in general use has the flux in cores that run through the length of the solder (Figure 9.14) and the size is 22 s.w.g. There are other solders that exist for a variety of temperatures and purposes. For example in response to the needs of a "greener" environment there is now a lead-free solder, and a flux that is halide free.

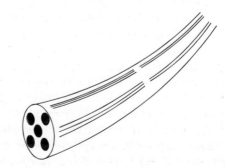

Figure 9.14   Five flux cores and solder

Soldering irons

The type and size of soldering iron (Figure 9.15) that should be used must be related to the connections that have to be made.

Figure 9.15

Some soldering irons operate at a fixed temperature and others can be variable. There are also different sizes and shapes for the bits which are interchangeable. For solder to flow on all of the surfaces that are being connected all the metal must reach a temperature of at least 190 °C. If there are large metal contacts that have to be heated then an iron capable of raising and maintaining the temperature must be used. It is very easy to apply too much heat especially where there are plastics in use. Heat transmitted through the metal will soon melt insulation or terminal bodies. An iron of suitable dimensions and temperature must be used.

Soldering techniques

There are two stages to making a good solder connection. First the separate parts of the joint must be prepared and secondly they need to be put together and soldered.

Preparation

The key to a well prepared connection is care and neatness. Care taken here can save time later.

There are two types of lead wires used:
- solid single strand
- multi stranded

Solid **single strand wire** should have any insulation removed to the required length and then the conductor is usually ready for connection.

A **coaxial cable** has inner and outer cylindrical conductors and tools are available for stripping the cable ready for connecting to the socket or plug (Figure 9.16). To connect the socket or plug unscrew it and slide the screw cap over the cable. First strip away 23 mm of the outer sheath and gather the strands of the copper screening, winding them back around the outer insulation. About half of the inner insulation should now be stripped and the screening wire should not touch the inner copper wire. Fit the claw, which clamps the outer screen, over the inner sheath and sit it around the screening (Figure 9.17). Push the inner copper wire through the hole in the remaining half of the plug and, when the screw cap is tightened, the outer screen should be trapped by the claws inside the plug. Any protruding wire should be cut off flush with the plug. Unless

you are using the "grub screw" version of the plug, for the best connection, the inner copper wire should be soldered.

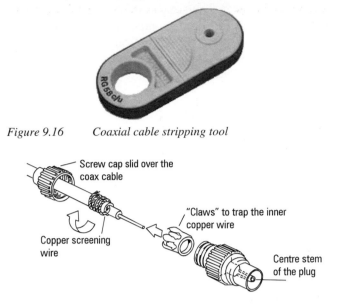

Figure 9.16   Coaxial cable stripping tool

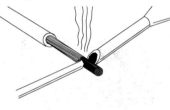

Figure 9.17   Coaxial cable and plug connection

When using a **stranded cable**, in addition to removing the insulation, the copper strands need to be twisted together and "tinned" (Figure 9.18). To tin the wire the soldering iron and solder are applied to the end of the twisted wire and then worked up towards the insulation. Tinning stranded conductors is a skill that needs practice for the amount of solder and heat used is critical. A good tinned wire should result in a bright shiny surface where all strands are joined together as one.

Figure 9.18

Part of the preparation of any joint is forming the lead wires to the required shape. There are several methods of soldering wires to solid terminations and we shall consider three of these in some detail.

Soldering to a pin

When connecting to a pin the wire should be bent to a hook shape using long nosed pliers, as shown in Figure 9.19. The hook should be placed over the pin and then squeezed to form a tight fit.

Figure 9.19

Soldering to tag strip

The connecting wire is threaded through the hole in the tag strip and then bent round the tag as shown in Figure 9.20. For components it may be more appropriate to bend the connecting wire 90° for ease of removal or replacement. Excess wire is then cut off and the joint squeezed flat with pliers. The tag strip is shown in Figure 9.21.

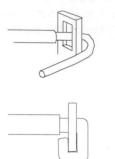

If components may need to be replaced it could be more appropriate to bend the connecting wire 90°.

Figure 9.20

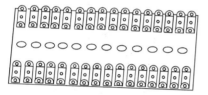

Figure 9.21   *Tag strip*

Soldering to printed circuit board

If it is a component to be fitted to the printed circuit board the leads must be formed to the correct pitch before being inserted. This can be carried out using a pair of long nosed pliers or special lead forming tools (Figure 9.22). The lead forming tool consists of two arms over which the component leads are bent. When a number of components all have to be prepared the lead forming tools ensures a consistent lead pitch on all components.

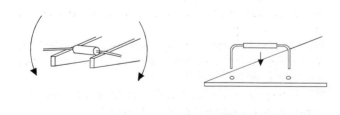

Figure 9.22

After the component has been inserted into the board the leads should be bent to keep the component in place (Figure 9.23). The leads can now be cut back leaving enough to hold the component but not too much so as to short out other tracks on the printed circuit board.

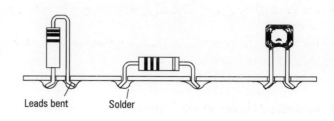

Leads bent   Solder

Figure 9.23

There are three main factors for a good connection.

• cleanliness
• correct heat
• correct flux (if a cored solder is used this should be ensured)

Remember cleanliness at the joint can be increased by tinning the surfaces first. That is to coat them in a thin layer of solder.

> ### Remember
> Aluminium conductors cannot generally be soldered in this way.

For good soldering joints

• the iron should be clean and fluxed
• when solder is applied to the iron it should appear bright and shiny if the iron is up to temperature
• the surfaces to be soldered should be clean and tinned
• the iron should be applied to the surfaces to be connected NOT to the solder
• solder should be seen to run over the surfaces when the joint is made
• the joint should be allowed to cool NOT cooled by blowing or applying a damp cloth

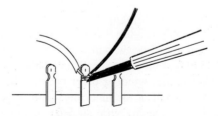

Figure 9.24

Precautions when soldering

- never flick off solder from the iron as this may cause harm to persons, clothing or apparatus
- where soldering is inside equipment it may be necessary to cover components
- heat sinks may need to be applied on some heat sensitive components
- always keep the soldering bit clean and tinned
- use the soldering iron stand to prevent danger from burns or fire.
- use a proprietary cleaner for removing flux residues so that the joint can be readily inspected

Common soldering faults

- **Dry joints** – These are caused generally by not using a high enough temperature or by moving the leads before the solder has "set". All surfaces have to get to a temperature high enough to melt the solder. A dry joint often looks grey and dull not bright and shiny as it should be.
- **High resistance joints** – dry joints are often also high resistance joints. It is not always possible to detect a high resistance joint visually. Where the surfaces have not been cleaned and tinned correctly it is possible for crystallisation to form between the surfaces, particularly if a solder is used without sufficient flux.
- **Excess solder** – can lead to connections being shorted out accidentally.
- **Too much heat** – can cause several problems including:
  - damage to insulation
  - damage to the printed circuit board
  - hardening of the wire making it brittle
  - damage to components - components may be protected by using a shield or a heat shunt.

A heat shunt (Figure 9.25) looks rather like tweezers and is clipped on the lead of a component before the lead is soldered to the circuit board. This diverts much of the heat travelling up the lead from the soldering iron and keeps the component relatively cool. If care is taken when soldering and the iron is only applied to the joint for a few seconds then a heat shunt is not necessary for most types of component, however they are advisable for germanium transistors or diodes.

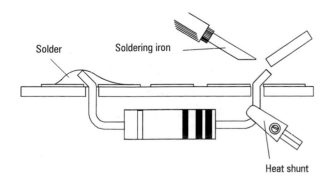
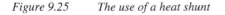

Figure 9.25   The use of a heat shunt

Desoldering techniques

Often when making or repairing electronic equipment it is necessary to remove a component from a printed circuit board. In order to avoid damage to the rest of the board cut away the component to be discarded before de-soldering. The technique of desoldering is used to remove the solder from the joint. Two methods are commonly used. The simplest to use is the desoldering braid (Figure 9.26). The braid is placed on the joint, then the soldering iron is placed on top to heat both the braid and the joint. When sufficiently heated the molten solder is wicked up into the braid.

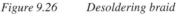

Figure 9.26   Desoldering braid

The second method is to use a desoldering tool. This has a spring loaded plunger that sucks hot solder from the joint when the plunger is released (Figure 9.27).

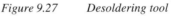

Figure 9.27   Desoldering tool

If excessive heat is applied when desoldering, the tracks can become detached from the board and broken. In addition to this, the components themselves can be damaged by heat.

Remember
The holes in the boards should be cleared and everything completely clean before mounting the new component.

Common desoldering problems

- When the desoldering tool is full of solder it does not operate smoothly.
- When the joint is not hot enough solder remains on the joint. If this happens it is often better to re-solder the joint before desoldering again.
- When the joint is too hot the PCB tracks may lift from the board or the components may be damaged.
- When the desolder nozzle does not seal to the board little suction is achieved.
- Some people leave the iron on the joint whilst applying the desolder nozzle – this can work with care but tends to damage the nozzle and the PCB as the desolder nozzle hammers the iron onto the PCB surface.

Electrical measurements

Taking measurements

When using test instruments take care not to use voltages and currents in excess of the component's rated value or the component may be destroyed. Before using the measuring instrument check that it has a current standard's certificate. This confirms that the calibration is correct.

Measuring resistance

The instrument used to measure resistance is an ohmmeter (Figure 9.28). As the values of resistance can range from a few ohms up to millions of ohms the ohmmeter must also have ranges capable of measuring this. Traditionally multi-range instruments have been used where the ranges are switched as required. More recently self ranging digital instruments have come into use. Whichever type of instrument is used it is important to have some idea as to what the resistance should be in order to check that the reading is correct.

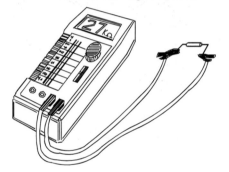

Figure 9.28   Measuring resistance

Before measuring resistance it is important to first check the condition of the instrument's internal battery, if this is low it can lead to inaccurate readings. Secondly check that the instrument has been set to zero with the leads connected together and the range switched to the one required.

Analogue readings

Measuring resistance with an instrument which has an analogue scale can sometimes be confusing. Figure 9.29 shows a typical resistance scale on a multimeter.

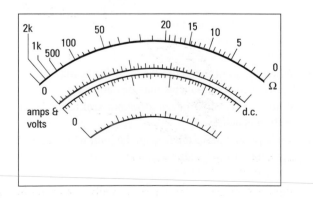

Figure 9.29   Analogue scale

The scale is not linear and the zero is at the opposite end of the meter to the other scales.

Testing diodes

A diode is a device which has a low resistance when a voltage is applied in the forward direction and a high resistance when applied in the reverse direction (Figures 9.30 and 9.31).

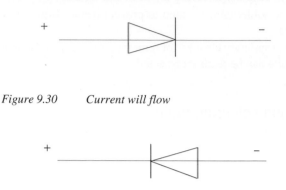

Figure 9.30   Current will flow

Figure 9.31   Current will not flow

Using the same instrument as used for measuring resistance it is possible to check if a diode is working correctly.

The ohmmeter has its own internal battery and this supplies the voltage to apply across the diode. The ohmmeter has its own polarity and this must be known to correctly apply the forward or reverse voltage (Figure 9.32).

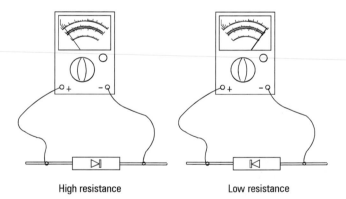

High resistance   Low resistance

Figure 9.32

> ### Remember
> On many analogue instruments, when using the resistance range, the negative terminal is more positive than the positive terminal. This means the instrument is connected to the diode in the opposite way than it would at first appear.
>
>

Digital instruments have a very low voltage output and unless they have a diode testing range they are not usually suitable for this purpose.

Testing transistors

As transistors consist of pnp or npn configurations (Figure 9.33) the testing of them is similar to diodes.

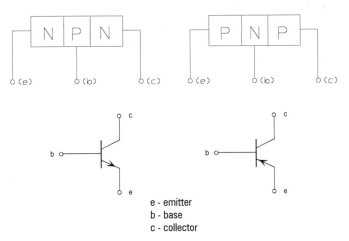

e - emitter
b - base
c - collector

Figure 9.33

There are special instruments with three terminals for checking transistors but an ohmmeter can be used for determining if a transistor is conducting correctly.

When connecting an ohmmeter to an npn transistor, as shown in Figure 9.34, the instrument should indicate a high resistance. Reversing the instrument connections to "b" and "c" should produce a low resistance.

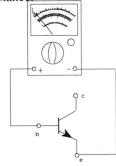

Figure 9.34

Similar results should be obtained if the instrument is connected between "c" and "e".

Testing a pnp transistor is similar but the low resistance readings will be obtained when the polarity is reverse to that of an npn.

The simple go/no-go transistor test may not be sufficient for proper diagnostic testing and a commercial test instrument may have to be used. This will mean that the transistor will have to be removed from the circuit but the test result will be far more comprehensive.

More complex devices such as integrated circuits are normally tested by replacement with another similar device. This should only be carried out if the circuit conditions are correct and in accordance with the manufacturer's specification otherwise the replacement device may well end up in the same damaged state as the original.

Measuring voltage

The instrument used for measuring voltage may be the same multimeter as used for testing resistance **but switched to the voltage range**. It may, however, be a completely independent voltmeter.

Whenever it is necessary to take voltage readings care must be taken as the readings have to be carried out when the circuit is live. To ensure it is not possible to get a shock safety precautions must be taken. If the voltage on the equipment you are working on exceeds 50 V a.c. or 120 V d.c. special test probes must be used which meet the requirements of the Health and Safety Executive Guidance Note GS38.

Voltage can be measured across a component or between any two points in a circuit (Figure 9.35). As with all testing it is important to have some idea what the reading should be. If this is not possible the instrument should be switched to a high range and then brought down to a suitable range which is appropriate to the readings obtained.

Figure 9.35   Measuring voltage drop

Measuring current

When an instrument is connected to measure current it should be in series with the load and the full current flows through the instrument. It is important that the instrument is suitable for the current that is being measured.

The supply should be switched off when the instrument is connected and disconnected. Often a circuit has to be broken to connect the instrument in series with the load. The circuit connections should be restored after the readings have been completed.

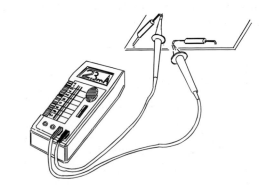

Figure 9.36   Measuring current

In Figure 9.36 the two resistors normally connected in series have been separated and the ammeter connected between them.

50
Testing components

Objective: To carry out tests on resistors, capacitors and semiconductor devices to verify the state of the component.

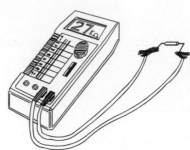

Suggested time: 1 hour

Materials:

• examples of resistors, capacitors, diodes, transistors

Procedures:

Using the ohmmeter or multimeter measure the resistance of a resistor, note the reading below, compare the reading to the colour code on the body of the resistor and verify the state of the resistor.

Remembering that for a polarised capacitor it is important to connect the true positive of the ohmmeter to the positive lead of the capacitor, and using the ohmmeter, measure the resistance of both a non-polarised and a polarised capacitor and note the readings below. If the resistance is less than about 1 MΩ then current is being allowed to pass from the ohmmeter and therefore the capacitor is leaking and faulty. Note that there may be an initial surge of current with large value capacitors (μF range). Verify the state of the capacitors in the space below.

Using the ohmmeter or the diode testing range in a multimeter measure the resistance of a diode, note the result below, reverse the component and make a note of the new result and verify the state of the diode.

Determine the emitter, base and collector pins of a transistor from the part number and manufacturer's data. Use the ohmmeter to verify the state of this component and make a note below of any readings taken.

Points to consider:

• was the condition of the instrument checked before tests were carried out?
• was the instrument set on an appropriate scale?
• was the polarity of the diode checked before testing?
• were the correct pins of the transistor identified?

Results:

Component	Results of test				Condition of component
Resistor	Colour code value		Reading		
Non-polarised capacitor	More than 1 MΩ?				
Polarised capacitor	More than 1 MΩ?				
Diode	Positive of instrument to anode test reading		Positive to cathode test reading		
Transistor (npn) e – emitter b – base c – collector	Positive to b, negative to c	Positive to b, negative to e	Negative to b, positive to c	Negative to b, positive to e	
Transistor (pnp) e – emitter b – base c – collector	Positive to b, negative to c	Positive to b, negative to e	Negative to b, positive to c	Negative to b, positive to e	

51
Basic soldering

Objective: To form good electrical connections by soldering.

Tasks:

Solder a length of 7/0.2 mm equipment wire, tinned stranded, to the matrix board copper track. Your tutor will indicate into which hole to solder the wire.

Press termination pins through the holes indicated by your tutor, solder the pins to the track and solder a length of 0.5 mm^2 PVC insulated stranded copper wire between the pins.

Press termination pins through holes indicated by your tutor, solder the pins to the track and solder a length of 30 A tinned fuse wire between the pins.

Press termination pins through holes indicated by your tutor, solder the pins to the track and solder a 1 W carbon resistor between the pins.

Solder a 0.6 W metal film resistor and two lengths of 0.5 mm^2 PVC insulated stranded copper wire to a tag strip as indicated by your tutor.

Suggested time: 1$\frac{1}{2}$ hours

Materials required:

suitable size piece of stripboard (1 mm holes) with enough room for the above tasks to be carried out
miniature tagboard
1 mm press-in pins
2   lengths of 0.5 mm^2/PVC insulated stranded copper wire
1   length of 7/0.2 mm equipment wire
1   length of 30 A tinned fuse wire
1   1 W carbon resistor
1   0.6 W metal film resistor

Procedures:

Strip the lead wires of insulation to the required length. Twist and tin the end of the stranded wire. Solder the wire and components into place,

Points to consider:

* were the wires terminated around the pins to an acceptable standard?
* were there any dry joints?
* was there an excess of solder on the joints?
* was there any excess wire at the termination?
* was the insulation damaged?

52
Soldering components to a circuit

Objective of task: To solder a variety of components to the stripboard so that they form a circuit as shown in the diagram.

Note: the circuit suggested can be used in further exercises in this book.

Suggested time: $1\frac{1}{2}$ hours

Materials:

copper track stripboard
terminal pins suitable to stripboard
resistors:

2×0.4 W 470 Ω
2×0.4 W 33k Ω

2×5 mm T $1\frac{3}{4}$ light emitting diodes
$2 \times$ BC184L transistors
$1 \times$ IN4002 diode
2×10 µF, 25 volts d.c. axial electrolytic capacitors
snap-on battery connector for PP6 battery
heat shunt

Procedures:

Connect the above components to the board as shown in the diagram. Use a heat shunt on the semi-conductors.

Points to consider:

• were all of the solder connections good?
• were any tracks shorted out by solder that should not be?
• were the components connected in the specified places?
• were the LEDs connected to the correct polarity?
• were the electrolytic capacitors connected to the correct polarity?
• were the transistors correctly mounted?

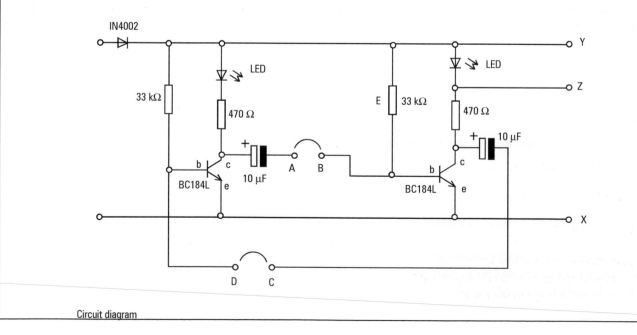

Circuit diagram

53
Replacing a component in a circuit

Objectives: To disconnect and replace a component on a circuit board. To complete a voltage test on the circuit.

Suggested time: ½ hour

Materials:

- completed previous exercise
- a 330 Ω resistor
- 2 lengths of tinned copper wire
- battery terminals and PP6 battery

To battery (black)

To battery (red)

Procedures:

Solder the battery connections as shown.
Connect the battery to the battery connectors and note the result.
Using the voltmeter test the circuit voltage at points X and Y
- negative to pin X and positive to pin Y. Record the results below.

Test the voltage at points X and Z - negative to pin X and positive to pin Z. Record the results. Typically the difference between the readings is approximately 2 V for red LEDs.

Remove the battery.

Solder in links between A and B, C and D.

Replace the battery and note the results.

Remove the battery. Remove component shown as resistor E on the above diagram and replace it with the 330 kΩ resistor. Check the soldering visually, replace the battery and note the result.

Results:

Voltmeter test at points X and Y
Voltmeter test at points X and Z

Difference between readings?

Points to consider:

- was there any excess solder left?
- was the new connection electrically sound?
- was there any damage to the track?

54
Terminating coaxial cable to a coaxial plug and socket

Objective: To terminate the coaxial cable into the plug and socket correctly to form a good electrical connection.

Suggested time: ¾ hour

Materials:

- a length of coaxial cable
- coaxial plug and socket, one solder version and one "grub screw" version

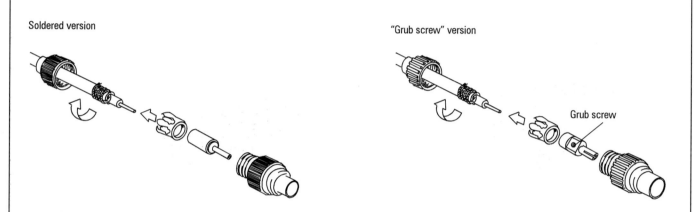

Soldered version

"Grub screw" version

Grub screw

Procedures:

Terminate each end of the coaxial cable onto the plug and socket. Using an ohmmeter check for a good electrical connection on the screen to screen and the core to core and confirm isolation between core and screen.

Points to consider:

- was the instrument checked before starting the test procedure?
- were the strands of copper screening turned back sufficiently to be clear of the core?
- was the cable correctly terminated to the plug or socket?
- was the connection electrically sound?
- in the soldered version has any damage been done to the plastic bush?
- was there any excess solder on the surface of the pin?

55
Terminating ribbon cable to an insulation displacement connector

Objective: To terminate the ribbon cable into connectors and test for continuity of each core.

Suggested time: ¾ hour

Materials (for this example):

- length of 10-way IDC standard flat ribbon cable
- 2 connectors – Universal style 10-way plugs conforming to BS9525-F0023
- Universal vice jaw tool common frame and insert appropriate for plug above

A vice will be required for this termination.

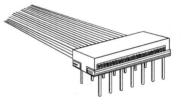

Procedure:

Terminate the ribbon cable into the connectors. Check for good electrical connections between each pin on connector 1 and the corresponding pin on connector 2. Confirm that no electrical connection exists between each pin on connector 1 and any other pin except its corresponding pin on connector 2.

Points to consider:

- was the instrument checked before starting the test procedure?
- is the connector assembled correctly?
- are all of the connections electrically sound?
- do the connections test out correctly?

Learning Resources
Centre